BOOKS BY PERCY A. MORRIS

WHAT SHELL IS THAT?
NATURE PHOTOGRAPHY AROUND THE YEAR
THEY HOP AND CRAWL
THE BOY'S BOOK OF SNAKES
A FIELD GUIDE TO THE SHELLS

A Field Guide to the Shells

of our Atlantic and Gulf Coasts

THE PETERSON FIELD GUIDE SERIES

A Field Guide to the Shells

of our Atlantic and Gulf Coasts

BY PERCY A. MORRIS

*Peabody Museum of Natural History
Yale University*

*Revised and Enlarged Edition
Illustrated with Photographs*

HOUGHTON MIFFLIN COMPANY BOSTON

ELEVENTH PRINTING   W

Editor's Note

THERE WAS a time, scarcely more than a generation ago, when many boys sparked their first interest in natural history by collecting bird's eggs. Eggs were a perfect outlet for that craving of all boys to collect things. They involved a kind of treasure hunt, were attractive in a gemlike way, and could be preserved. With some lads the hobby never became more than a mild form of kleptomania; other boys, more thoughtful, eventually became deeply interested in the birds themselves. Many top ornithologists among the older men can trace the genesis of their science to early collecting. A few of them deplore the fact that egging is outlawed today, but in the interests of conservation it is best that it is no longer allowed.

Although butterflies have their points and so do minerals, no natural history objects lend themselves better to the gratification of this collecting instinct than shells. Many collectors of course are interested in shells in the same way that one would be in jewels, jades, or fine porcelain. Others, aware that the shell is but the garment of a once living animal, take the naturalist's point of view and concern themselves with classification and distribution. All too few inquire into the life of the living mollusk, its ecology and its habits.

Recognition, however, always comes first. That is why the Field Guide Series was launched — as a short cut to recognizing and naming the multitude of living things which populate America. The first volume to appear, *A Field Guide to the Birds*, met with instant success; then followed *A Field Guide to Western Birds*, and now Percy Morris's admirable *A Field Guide to the Shells*.

Whether your interest in the seashore is that of a bird watcher (no seascape is ever devoid of birds), that of a surf fisherman casting for channel bass, or merely that of a beach comber, you cannot overlook the shells. From the rocky headlands of Maine to the long sandy beaches of the Texas coast stretch thousands of miles of good collecting grounds. Rock, sand, and mud offer radically different environments. Some specimens you will find at high-tide line where the sea has cast them up. Others will be in the tide pools. Still others you must grapple for in deep water.

Take this book with you whenever you go to the shore. Do not leave it home on your library shelf; it is a Field Guide, meant to be used.

ROGER TORY PETERSON

Preface to the Second Edition

IN THIS, the second edition of *A Field Guide to the Shells of our Atlantic and Gulf Coasts*, one hundred and twelve species of mollusks not covered in the first edition are described and illustrated. There are six new plates of photographs, and in addition several of the earlier plates have been rebuilt, with new photographs. An attempt has been made to include the common, or popular, name wherever such a name has become established, and in each case the author of the species is given. Following the descriptions of Families, the date of erection for each Genus is given, along with the number of species or varieties within that genus that are recognized as occurring in the area covered by the book. In addition, much of the text has been rewritten, and the nomenclature brought up to date. It is hoped that this new edition will prove even more useful to the growing army of enthusiasts who are embracing shell-collecting as a hobby.

P. A. M.

NEW HAVEN, CONNECTICUT
Nov. 14, 1950

Preface

THE FOLLOWING is an attempt to present, in non-technical language, descriptions of the marine shells that may be collected along the East coast, from Maine to Florida. The total number of species known to exist along our Eastern seaboard runs into the thousands, and it would be manifestly impossible to discuss them all in the pages of an ordinary-sized book. The selection here presented is believed to include all of the larger and commoner forms, with a liberal sprinkling of the minute, rare, and deep-water varieties, so that the seashore visitor should be able to identify almost any molluscan shell picked up on our Atlantic coast.

The author wishes to acknowledge his indebtedness to Professors Carl O. Dunbar and Stanley C. Ball, of Peabody Museum, Yale University, for their constructive criticisms and advice, and especial thanks are due Doctor William J. Clench, of Harvard University, for generously giving of his time in reading the manuscript, and for many suggestions regarding matters of nomenclature. While these gentlemen aided no little in the final preparation of this book, the author accepts full responsibility for any errors that may creep into its pages.

P. A. M.

NEW HAVEN, CONNECTICUT
May 11, 1946

Contents

THE PELECYPODS

THE GASTROPODS

List of Illustrations

Introduction

SHELL-COLLECTING is one of the oldest of natural history hobbies. This active sport has been enjoyed by peoples from the earliest days of recorded history, and from necklaces and other adornments found with ancient Indian burials we know that even savage man appreciated the beauty of shells.

A century ago the sport was so firmly established that regular auctions were held in the larger cities, and ardent collectors came from near and far to bid on the choice and colorful shells, while scores of dealers made comfortable livings supplying the demand for rare and unusual specimens. The hobby, long dormant, appears to be reviving considerably of late, and while it may never approach the popularity it enjoyed in its heyday, it still offers much to the devotee. It is a healthy, outdoor pursuit, and a collection, properly arranged, makes an extremely interesting and colorful display. Our butterflies and moths are often considered the last word in brilliancy of colors and delicacy of shades, but they find worthy rivals in many of the tropical mollusks.

A collection is quite permanent, as the shells do not lose their colors readily, are not subject to decay, and are not attacked by injurious insects. Many of our leading naturalists acquired a taste for the pursuit of natural history, when but youths, in collecting and studying shells.

One scarcely needs to be told where to collect specimens. They occur from the depths of the ocean to the limits of vegetation on mountain tops. Mollusks abound in fresh-water streams and ponds, and are perhaps even more numerous in stagnant swamps and ditches. Terrestrial varieties will be found in woods and rich meadows, commonly under logs and stones, or under bark.

Marine mollusks, the ones we are concerned with in this book, are found under a wide variety of conditions. Some species favor muddy shores, some sandy stretches, and some prefer rocky coasts. Some demand deep water while others prefer shallow water. Many insist on pure, clean sea water, while a few seem to get along better in brackish areas. Search everywhere, in rock crevices and on open beaches. Tide pools, left like miniature aquaria by the receding tides, are often real treasure-troves, and the line of debris, called 'sea-wrack,' that marks the limit of high water, will repay a careful going over. Any stone or boulder that is not too heavy should be overturned, as a host of different creatures take advantage of such shade and protection while waiting for the return of the next tide. Provide yourself with a stout shovel and dig deep into the sands, muds, and gravels near the low-water level, and you will be rewarded by finding many shells that you would not otherwise see.

Many of the shells found will be empty, but should the animal be present, it can be removed by immersing the shell in moderately hot water for a few minutes. A hairpin bent to form a hook will aid in the process. If the soft parts break off, leaving some in the shell, simply place the whole thing in alcohol for a few days and dry in the shade, after which there will be no odor. Very minute shells are best so treated.

The epidermis of shells may be removed by soaking in a solution of caustic soda, about one pound to a gallon of water. This solution should be handled carefully, as it will eat your own epidermis as well as that of the shells. While this process results in decorative shells, most collectors prefer their specimens natural.

A collection's value is proportional to its labeling. The precise locality and date are far more important than having the shell correctly named. Adopt some form of label that provides space for a number (which will be entered in your catalogue and placed, in waterproof ink, on each specimen), the name, date, and locality. This last should be as exact as possible. 'East coast of Florida' does not mean very much, but '1.5 miles north of Miami, Florida' ties the specimen down accurately. The collector's name should be included, as well as the name (or initials) of the person responsible for its identification. In the catalogue, under that number, you may include notes on the tide conditions, weather, whether or not the mollusk was a living specimen or an empty shell, and anything else that may seem pertinent. Below is a sample of the kind of label used by the writer.

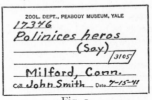

ZOOL. DEPT., PEABODY MUSEUM, YALE

17376

Polinices heros

(Say) /3105/

Milford, Conn.

Coll. *John Smith* Date *7-15-41*

Fig. 1

In order to properly understand any group of objects, some system of classification, or orderly arrangement associating those of a kind, is necessary. To appreciate the necessity of such systematic arrangement, we have only to imagine trying to use a telephone directory in which the names were not arranged in alphabetical order.

Animals differ very much from one another, not only in their form, size, and habits, but also in their internal structure. If we were to consider external form only, we would have to include the whales with the fishes, and group such totally unrelated animals as snakes, eels, and earthworms.

The accompanying chart will perhaps explain better than written words the system used in modern classification. There are three Kingdoms — Vegetable, Animal, and Mineral. The Animal Kingdom is split into about a dozen major divisions known as *Phyla*, a term derived from the Greek word meaning a tribe. Each *Phylum* consists of a group of animals that are alike in some fundamental structural character.

For example, the *Phylum Chordata* contains all of the animals that possess a backbone, while all of the animals with external skeletons and jointed legs belong to the *Phylum Arthropoda*. The *Phylum Mollusca*, the one we are concerned with, is made up of creatures which have a soft, unsegmented body, which may or may not be protected by a limy, external shell. The next grouping below *Phyla* is into classes, and the mollusks fall naturally into five, of which we are particularly interested in the third and fourth.

Class 1   Amphineura — the chitons
Class 2   Scaphopoda — the tooth-shells
Class 3   Pelecypoda — the clams
Class 4   Gastropoda — the snails
Class 5   Cephalopoda — the squids, etc.

If we were to take the *Phylum Chordata*, the next division also results in five classes, well known to all of us, the birds, mammals, fish, reptiles, and amphibians. The next step is to separate those of each class into orders, and, selecting the mammals, we divide them into carnivores, rodents, etc., each step giving us more restricted groups. If we concentrate on a certain snail, we place it first in the order containing most of the marine snails, and from there move it to the family it most closely resembles. The next step is to the genera, and we find our shell is obviously a *Busycon*. There are four members of this genus living on our East Coast, and our specimen fits the description of *Busycon carica*. If our shell happens to be exceptionally rugged and heavy, with a swollen canal, it is probably *Busycon carica elecians*, a subspecies. Thus we have the following order, Kingdom — Phyla — Class — Order — Family — Genus — Species — Subspecies. (See p. 3.)

The beginner is apt to be confused and, at first, exasperated by some of the 'jaw-breaking' names that have been given to living creatures. This is true in any branch of natural history, but is perhaps more apparent in the molluscan field, since only a few kinds have become well enough known by the layman to have acquired 'popular' names. Perhaps this is a good thing, as English names are very likely to be too local in character. Thus the same shell may be called one thing in one place, and go under a completely different name a few miles down the coast, while in three widely separated localities the same name may be applied to three totally different shells.

Scientists have agreed upon Latin for the naming of animals (and plants) for very good reasons. It is a dead language, and therefore not subject to change. Serious scholars all over the world, regardless of their individual nationalities, are familiar with it, so that it is about as close as one can get to an international tongue. The name 'round clam' means something to a Carolinian, while 'quahog' means the same thing to a New Englander, but neither means anything to a Frenchman or a Russian. Use the bivalve's scientific name of *Venus*, however, and every shell enthusiast, whatever his country, knows what shell we are talking about.

In this book the popular names have been given wherever possible, but the serious collector is strongly urged to learn to recognize his treasures by their authentic names. Scientific names are not difficult after we become familiar with them. We use the words alligator, boa-constrictor, and gorilla commonly enough, without stopping to realize that they are perfectly good scientific names. Especially in the field of flowers do we all rattle off scientific names at a great rate, talking about our crocus, iris, delphinium, chrysanthemum, wistaria, forsythia, geranium, cosmos, and countless others with the greatest of ease. There are many common English words, such as secluded and strategic, that are much harder to pronounce than *Mya*, *Conus*, or *Cardita*.

It has been found convenient to give each animal two names; first a general, or generic, name (always capitalized), which indicates the group (genus) to which it belongs, and second, a special, or specific, name (never capitalized), to apply to that animal alone.

Thus we have, for example, a well-known group of marine clams incorporated in the genus *Venus*, so named because of their beauty of form and symmetry. The shells of one of them, the familiar 'round clam' referred to above, were used by the coast Indians for money, so that particular species was named *mercenaria* by the great Swedish naturalist, Linnaeus, in 1758.

The name of the person who first describes a species (known as the author) follows the scientific name. Thus our round clam becomes *Venus mercenaria* Linne. If later research proves a species belongs in a genus different from the one to which the original author has allocated it, it is removed to the proper genus, but the species name is retained, and parenthesis are placed around the author's name to indicate that there has been a change from the original description. This latter is a technicality of importance chiefly to specialists, and this book follows the custom of other popular publications in not using the parenthesis. Varieties, or subspecies, differing slightly but constantly from the types, but not sufficiently to accord them full specific rank, are indicated by a third name. A smaller and more colorful round clam of our southern shores, described by Thomas Say, is known as *Venus mercenaria notata* Say.

The *Phylum Mollusca* forms one of the major branches of the Animal Kingdom, including as it does the clams, oysters, and scallops, the snails and slugs, and the chitons, squids, octopi, and some others. Some seventy-five thousand living kinds are known throughout the world, besides many thousands of fossil species, for the group is represented in the most ancient of fossil-bearing rocks. In this book we are concerned only with the clams and snails occurring on the eastern coast of North America, from Labrador to Texas, but at this point a very brief discussion of the other three classes might be in order.

The *Amphineura* are primitive mollusks of very sluggish habits, mostly preferring shallow water close to shore. The typical chiton is an elongate, depressed mollusk, bearing a shelly armor of eight saddle-shaped plates arranged in an overlapping series along the back. The 'foot' is broad and flat, and serves either as a creeping sole or a sucking pad by which the creature clings to the rocks. The ancestral mollusk, from which all existing forms have evolved, is believed to have been very similar to the present-day chiton. A rather common species from Florida is illustrated on page 32.

The *Scaphopoda* are elongate mollusks enclosed in a tapering, conical shell that is open at both ends and slightly arched. From the larger end project the 'foot' and several slender tentacles. Scaphopods live in clean sand from shallow water to considerable depths, and their shells look like miniature elephant tusks. They are commonly called 'tusk shells,' or 'tooth shells.' An example is shown on page 32.

The *Cephalopoda* are highly specialized mollusks, keen of vision and swift in action. The head is armed with a sharp, parrot-like beak, and is surrounded by long, flexible tentacles that are studded with sucking disks. Besides the well-known devil-fishes (octopi), this class includes the delicate paper argonaut and the beautiful pearly nautilus of Asiatic waters. On page 32 is shown the internal 'pen' (a vestigial shell) of one of the common squids, an example of the nautilus and the argonaut, and the internal, chambered shell of a floating cephalopod, *Spirula*. These latter coiled shells, about an inch in diameter, are frequently found on southern beaches, but the living mollusk is not often seen.

In the Appendix there is a glossary of terms used by conchologists, some of which are used throughout this book. On page 4 will be found a plate showing the important parts of a snail shell, and of a clam shell. A brief study of this plate will help in reading the descriptions of the various species.

Part One

The Pelecypods

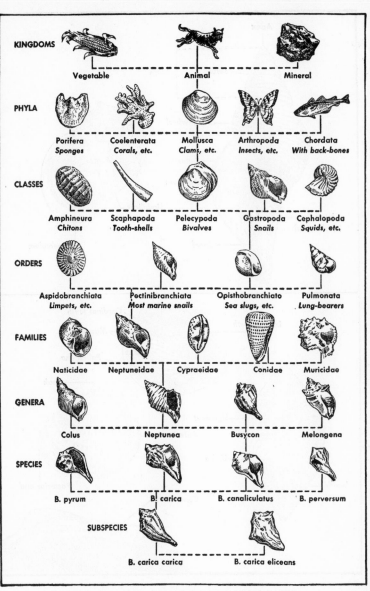

KINGDOMS
Vegetable   Animal   Mineral

PHYLA
Porifera *Sponges*   Coelenterata *Corals, etc.*   Mollusca *Clams, etc.*   Arthropoda *Insects, etc.*   Chordata *With back-bones*

CLASSES
Amphineura *Chitons*   Scaphopoda *Tooth-shells*   Pelecypoda *Bivalves*   Gastropoda *Snails*   Cephalopoda *Squids, etc.*

ORDERS
Aspidobranchiata *Limpets, etc.*   Pectinibranchiata *Most marine snails*   Opisthobranchiato *Sea slugs, etc.*   Pulmonata *Lung-bearers*

FAMILIES
Naticidae   Neptuneidae   Cypraeidae   Conidae   Muricidae

GENERA
Colus   Neptunea   Busycon   Melongena

SPECIES
B. pyrum   B. carica   B. canaliculatus   B. perversum

SUBSPECIES
B. carica carica   B. carica eliceans

Modified from an exhibit at the Cranbrook Institute of Science, Bloomfield Hills, Michigan

CLASSIFICATION CHART FOR THE MOLLUSKS

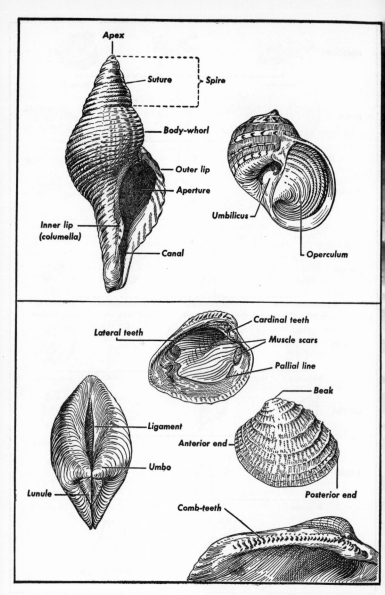

TERMINOLOGY OF SNAIL AND BIVALVE SHELLS

Family Solemyidæ

IN THIS FAMILY the valves are equal, and considerably elongated. There is a tough and glossy periostracum that extends well beyond the margins of the shell. These are rather uncommon bivalves, but the family is represented on our east coast from Canada to the West Indies.

Genus *Solemya* Lamarck 1818

4 Species

SOLEMYA VELUM Say   (Awning Shell)   p. 33

This is an exceedingly delicate little shell, from an inch to an inch and a half in length, found from Nova Scotia to Florida. It is somewhat oblong in shape, with beaks that are scarcely elevated at all. The surface is smooth, with about fifteen slightly impressed double lines, most conspicuous on the posterior margin. The chestnut-brown periostracum is firm, elastic, and shiny, and it projects far beyond the shell itself, where it is slit at each of the radiating lines, giving the edges a ragged, fringed appearance. The interior of the shell is bluish white, and its substance is so thin that the radiating lines are quite visible within. Known by such popular names as awning shell, veiled clam, and swimming clam, this bivalve is very easily recognized by the tough covering that overhangs the edge of its shell like a veil. It lives in the muds and sands below the low-water line, and appears to be most abundant during the spring months, in some years occurring in considerable numbers. *Solemya borealis* Totten is a larger species, averaging about twice the size of the more familiar *velum*. Its range is more to the north, from Massachusetts to Labrador. The color is deep tan or greenish brown, with the interior lead-color. Examples of *Solemya* are exceedingly fragile when dry, and should not be kept loose in trays. They should be attached by a drop of adhesive to a card cut to fit the tray, or they should rest upon soft cotton. Specimens that are free to shift about as a drawer is opened and closed soon become mere shell fragments.

Family Nuculidæ

SHELLS THREE-CORNERED or oval, and small, with no pit for the ligament between the umbones. There is a series of tiny but distinct teeth on each side of the beak cavity. The inside is polished, often pearly, and the inner margins are crenulate. Distributed in nearly all seas, but commonest in cool waters.

Genus *Nucula* Lamarck 1822

16 Species and varieties

NUCULA PROXIMA Say (Near Nut Shell) p. 33
The near nut shell is a tiny bivalve that lives by the thousands in sheltered harbors along the Atlantic coast from Maine to Florida. Only a little more than one-fourth inch long, and white in color, it has a thin, olive-brown periostracum that is commonly missing on beach specimens. The shell is triangular, the beaks somewhat elevated and inclined forward. The surface is marked longitudinally with scarcely perceptible striæ. The interior is pearly and highly polished, with the margins crenulate. The teeth of the hinge, while diminutive, are sturdy, the posterior series very distinct and regular. Examples of the near nut shell can usually be found among the beach litter on suitable shores. It frequents sandy or pebbly bottoms in moderately shallow water and is a very common shell in the stomachs of marine fishes. It is also eaten by various bottom-feeding ducks.

NUCULA TENUIS Montagu (Thin Nut Shell) p. 40
This little shell is much like the last, but the posterior margin is rounded rather than pointed, so that the outline is more oval than triangular. The color is pale yellowish green, with the interior silvery white, but not pearly. The shell is about one-fourth inch long, and quite thin and fragile. The hinge teeth usually number four or five before the beaks, and about eight behind. This species is not as abundant as the last. It lives in rather deep water, from Labrador to North Carolina, and has been taken in the Florida Straits.

NUCULA VERRILLII Dall (Verrill's Nut Shell) p. 40
This is a rare shell, living in deep water from Massachusetts to Mexico. Its length is less than one-fourth inch, and its shape is sharply triangular, with the beaks pointed. The color is pale brown, with the interior glossy white but not pearly. This little clam was discovered by Professor Verrill, of Yale, in 1885, and named *trigona*. That name, however, proved to be pre-empted, and in 1886 Dall renamed the bivalve *verrillii* in honor of the distinguished Yale zoologist.

Family Nuculanidæ

THESE WERE FORMERLY classed with the Nuculidæ. The shell is more or less oblong, usually rounded in front and prolonged into an angle behind. The margins are not crenulate. There are two lines of hinge teeth, separated by an oblique pit for the ligament. Distributed widely, but chiefly in cool seas.

Genus *Nuculana* Link 1807

19 Species and varieties

NUCULANA ACUTA Conrad   (Pointed Nut Shell)   p. 33

The pointed nut shell is about one-half inch long. The color is white, with a very thin brownish or greenish periostracum. In this bivalve the anterior end is broadly rounded, while the posterior end is prolonged to an acute tip. The shell is sculptured with distinct concentric grooves. The hinge is composed of a number of V-shaped teeth, interrupted by a central pit for the ligament. The interior is highly polished. This little clam occurs from Massachusetts to the West Indies, living in sandy mud a few feet beyond the low-water line. Empty valves are not uncommon in the litter on many southern shores, this species being perhaps less abundant in the North. Living examples can generally be obtained by sifting the bottom material just offshore.

NUCULANA CONCENTRICA Say   (Concentric-lined Nut)
p. 192

This species is roughly one-half the size of the last, being about one-quarter of an inch long. The shape is about the same, with the posterior tip less acute. The surface is decorated with sharply defined concentric grooves, and the color is white. The interior is polished. This species is at home in the Gulf of Mexico, occurring from Florida to Texas.

NUCULANA TENUISULCATA Couthouy   (Sulcate Nut)
p. 33

This is an oddly shaped little pelecypod. Averaging about one-half inch in length, and pale greenish gray in color, the shell is quite thin, the valves scarcely inflated at all. The posterior part is double the length of the anterior, and is narrowed to a truncate tip. The surface bears strong concentric grooves that are closely spaced. This little fellow is not often seen on the beach, but individuals may sometimes be obtained by dredging in the soft mud well out beyond the lowest tide limits. Specimens are commonly found in the stomachs of cod. This is a northern species, ranging down the coast as far as southern New England.

NUCULANA BUCCATA Steenstrup   p. 192

This is a stout little shell about three-fourths of an inch long. It is nearly as high as the last species, but is shorter, and less pointed, or pinched out, at the posterior end. The beaks are moderately prominent, and the hinge teeth are robust. The shell is fairly thick. The sculpture consists of closely spaced concentric lines, and the color is pale brown. This is a circumpolar pelecypod, living on our coast from Greenland to the Gulf of St. Lawrence.

NUCULANA CARPENTERI Dall   (Carpenter's Nut Shell)

p. 192

This is a narrow, elongate shell slightly less than one-half inch in length. The anterior end is broadly rounded, and the posterior end is drawn out to a small square tip. The dorsal line is considerably concave, with the beaks about one-third closer to the front end. The hinge teeth are delicate and few in number. The color is greenish gray, with the surface quite shiny, and there are very faint concentric lines. This is a graceful little bivalve, considered a prize in most collections. It lives from North Carolina to the West Indies, and prefers deep water.

Genus *Yoldia* Moller 1842

28 Species and varieties

YOLDIA LIMATULA Say   (File Yoldia)   p. 12

The file yoldia, so called for the filelike arrangement of teeth along the hinge line, is a handsome, shiny, deep-green clam, two to two and a half inches long. The shell is oval, much elongated, and thin, with the beaks nearly central but not prominent. The anterior and basal margins are regularly rounded, with the posterior end drawn out to a pointed and partially recurved tip. The interior is bluish white, the teeth along the hinge line being quite prominent, numbering about twenty on each side. This "streamlined" bivalve may be distinguished by its length, which is more than twice as great as its height, and by its peculiarly upturned, snoutlike posterior end. It is an active mollusk, swimming capably, and possessing the ability of leaping to an astonishing height. It has an extended range, being found on both sides of the Atlantic, and on the Pacific coast as well.

YOLDIA SAPOTILLA Gould   p. 33

This little yoldia is a smaller clam, seldom exceeding one inch in length. The shell is rather oval, somewhat elongated, thin and fragile, and green in color. The anterior end is regularly rounded, the posterior end narrowed, the line running from the beaks to the tip being nearly straight. The interior is white, with about fifty teeth on each side of the beak cavity. This little pelecypod, less elongate than the last species, is very commonly found in the stomachs of fishes taken off the New England coast. It lives in the mud of moderately deep water, but single valves are often to be seen among the beach litter that marks the high-tide limit. It may be collected from Nova Scotia to Cape Cod.

YOLDIA MYALIS Couthouy   p. 192
This is a smooth, elongate-oval shell averaging slightly less than one inch in length. The beaks are low, and situated close to the middle of the shell. Both ends are rounded, with the posterior end a little narrower than the anterior. There are about a dozen teeth in each series. The color is yellowish olive, and there is a thin periostracum that in fresh specimens is often arranged in alternate darker and lighter zones. The interior is glossy yellowish white. This species may be found from Labrador to Massachusetts. On our west coast it ranges from Puget Sound north.

YOLDIA ARCTICA Gray   (Arctic Yoldia)   p. 192
This is a rather squarish, or oblong-oval shell, well inflated, and attaining a length of nearly one inch. The beaks are quite prominent, and are located about one-third closer to the anterior end, which is regularly rounded. The posterior end is truncate, the basal margin sloping upward to form a blunt point. There are from twelve to fourteen teeth in each series. The color is lead-gray, with a thin greenish periostracum, and the interior is white. This species is a lover of cold water. Fossil shells (Pleistocene) are to be found in New Brunswick and Maine, and living examples occur from the Gulf of St. Lawrence north to Greenland.

YOLDIA THRACIÆFORMIS Storer   (Broad Yoldia)
                                                                p. 33
The broad yoldia is a substantial, squarish shell, strong and firm, broadest behind, and gaping at both ends. The general shape strongly suggests the blade of an axe. There is an oblique fold extending from the beak to the posterior third of the basal margin, giving the exterior a wavy appearance. The color is brownish olive, the inside pure white, not shiny. The hinge has a spoon-shaped cavity for the ligament, with about twelve robust teeth on each side. This is a fair-sized clam, growing to a length of nearly three inches. With its squarish outline, wavy surface, and salient rows of V-shaped teeth along the hinge line, it is not likely to be confused with any other shell. It lives in rather deep water, and is not uncommon on the New England coast.

Genus *Malletia* Desmoulins 1832

6 Species

MALLETIA OBTUSA Sars   p. 192
This shell is rather oblong in shape, and about three-fourths of an inch long. The beaks are very small. The anterior end is

short and evenly rounded, while the longer posterior end is abruptly rounded. The dorsal line from the beaks to the posterior tip is straight, with the line in front of the beaks gradually curved. There are numerous teeth in the hinge line, about sixteen in the front series and more than thirty in the rear. The color is greenish yellow in live specimens, pale straw-color with dead shells. This species prefers deep water, and ranges from North Carolina to Massachusetts.

MALLETIA DILATATA Philippi   p. 192
Like the last shell, this one does not seem to have any common name. It is a very neat little fellow, nearly one-half inch in length. The valves are inflated, so that the shape is almost box-like, and the beaks are somewhat swollen. The anterior end is acutely rounded, while the longer posterior end is abruptly truncated. There are about ten strong teeth in the forward series, and about sixteen in the rear. The color is a glistening white, inside and outside, and there is a sculpture of narrow but distinct concentric lines. This little bivalve is also found in deep water, off the coast of southern Florida.

Genus *Tindaria* Bellardi 1875

6 Species

TINDARIA AMABILIS Dall   p. 40
This is a solid little shell, about one-half inch long, and rather plump in outline. The shape is oval, with both ends rounded, the posterior slope a little longer and more pronounced. There are about a dozen teeth in the anterior series and nearly thirty in the posterior. These teeth are robust at the two ends, and grade gradually to very small teeth under the beaks. The valves are sculptured with strong concentric grooves, and the color is white, with a dull yellowish periostracum. The interior is polished white. This species lives in deep water, off our southern shores.

Family Arcidæ

IN THIS GROUP the shell is rigid, strongly ribbed or cancellated, with the hinge line bearing numerous comblike teeth arranged in a line on both valves. There is usually a heavy, commonly bristly periostracum. There is no siphon. Some members of this group crawl about in sand or mud, some prefer to attach themselves to objects by a silky byssus. They are world-wide in distribution, from shallow water to considerable depths (300 fathoms).

Genus *Arca* Linne 1758

15 Species and varieties

ARCA PEXATA Say   (Bloody Clam)   p. 33
This is a very common and well-known member of the family Arcidæ. It occurs from Maine to Florida, and gets its popular name from the fact that it is one of the very few mollusks having red blood. About two inches long, the shell is thick and solid, with prominent beaks that terminate in points that are nearly in contact, so that when viewed from the end the shell is heart-shaped. The surface has about thirty-five radiating ribs, somewhat broader than the spaces between them. The color is white, but the lower two-thirds of the shell is covered with a thick and shaggy greenish-brown periostracum. This covering does not usually remain on the shell for long after death, and the majority of valves picked up on the beach are a dull lusterless white. This species thrives on sandy bottoms in relatively shallow water.

ARCA AURICULATA Lamarck   (Eared Ark)   p. 40
This is a thick, heavy, and sturdy shell, rather oblong in shape, attaining a length of about two and one-half inches. The beaks are well elevated and directed forward, with the anterior end short and rounded and the posterior long and squarish. The hinge teeth are numerous and small. The surface bears about twenty-five strong radiating ribs which are crossed by threadlike lines. The color is white, with a silky brown periostracum. Known as the eared ark from the thin dorsal edge of the posterior tip, this fine shell occurs in southern Florida and the West Indies.

ARCA TRANSVERSA Say   (Transverse Ark)   p. 40
The transverse ark is only about one inch in length. Dull white in color, the shell is transversely oblong, with prominent teeth along the hinge line. The beaks are incurved, and placed at the anterior third of the shell. There are about thirty-two ribs, broader at the margins and narrowing toward the beaks. The two valves are slightly unequal, so that the margin of one passes a little beyond that of the other. The hairy periostracum is dark chestnut-brown. This species is easily identified. It lives on sandy or muddy bottoms just offshore, and its valves are generally common in the beach litter from New England to Florida.

ARCA INCONGRUA Say   (Incongruous Ark)   p. 40
This is a southern shell, ranging from the vicinity of the Carolinas to Texas. About two inches long, the white shell is inequivalve, rather short for its height, and considerably inflated.

MISCELLANEOUS PELECYPODS

All shells approximately one-half natural size.

1. *Donax variabilis* 4 views: **COQUINA**   p. 81
   Small; wedge-shaped; vari-colored.

2. *Arca occidentalis* 1 view: **TURKEY WING**   p. 14
   Wavy brownish streaks; long line of hinge-teeth.

3. *Glycymeris americanus* 1 view: **AMERICAN BITTERSWEET**
   p. 17
   Two rows of curved hinge-teeth.

4. *Dosinia elegans* 1 view: **ELEGANT DISK**   p. 66
   Orbicular; white; concentric lines.

5. *Yoldia limatula* 1 view: **FILE YOLDIA**   p. 8
   Thin-shelled, pointed; shiny green.

6. *Spondylus americanus* 1 view: **SPINY OYSTER**   p. 22
   Interlocking hinge; very spiny.

7. *Trigoniocardia medium* 1 view: **OBLIQUE COCKLE**   p. 64
   Pronounced posterior slope.

8. *Mytilus recurvus* 1 view: **BENT MUSSEL**   p. 35
   Kidney-shaped; strongly curved; radiating lines.

9. *Lævicardium lævigatum* 1 view: **EGG-SHELL COCKLE**
   p. 65
   Thin-shelled but strong; polished.

10. *Divaricella quadrisulcata* 1 view   p. 57
    Small; white; peculiar "bent" sculpture.

11. *Anatina canaliculata* 1 view: **CHANNELED DUCK**   p. 88
    Thin shelled; inflated; widely spaced concentric ribs.

12. *Dinocardium robustum* 1 view: **GREAT HEART**   p. 62
    Our largest cockle; brown area on posterior slope.

13. *Trachycardium muricatum* 1 view: **COMMON COCKLE**
    p. 62
    Small sharp scales on ribs; interior yellowish.

14. *Echinochama arcinella* 2 views: **SPINY CHAMA**   p. 55
    Thick and solid; white; spiny.

15. *Chama macerophylla* 1 view: **JEWEL BOX**   p. 55
    Heavy, thick, irregular; usually pinkish.

Plate 2   13

MISCELLANEOUS PELECYPODS

All shells approximately one-half natural size.

1. *Pecten gibbus* 4 views: **CALICO SCALLOP**   p. 24
   Small, vari-colored; wings about equal.

2. *Pecten muscosus* 1 view: **ROUGH SCALLOP**   p. 24
   Sharp scales on ribs; commonly reddish.

3. *Lima scabra* 1 view: **ROUGH FILE SHELL**   p. 30
   White or yellowish; ribs studded with sharp scales.

4. *Pecten irradians* 1 view: **BAY SCALLOP**   p. 23
   Ribs broad and rounded; wings slightly unequal.

5. *Pecten ornatus* 1 view: **ORNATE SCALLOP**   p. 25
   Wings unequal; a few ribs unspotted.

6. *Pecten sentis* 1 view: **THORNY SCALLOP**   p. 25
   Wings unequal; small; commonly reddish.

7. *Pitar fulminata* 2 views: **LIGHTNING VENUS**   p. 67
   Small, smooth; brown radiating lines.

8. *Pecten nodosus* 1 view: **LION'S–PAW**   p. 26
   Large; reddish; blunt knobs on ribs.

9. *Anomalocardia   brasiliana*   2   views: **LITTLE   STRIPED
   VENUS**   p. 71
   Prolonged posterior end.

The beaks are high, and are not well separated, so that in many instances you will find specimens with the beaks showing signs of wear, often being abruptly flattened. There are about twenty-five broad ribs, with narrow grooves between them, and these ribs are crossed by equidistant lines, less conspicuous toward the posterior end. The row of comblike teeth is graduated, with the smaller ones at the center. Maxwell Smith has pointed out the interesting fact that when looking directly at the basal margin, with the beaks directed away from the viewer, that margin presents a gentle curve instead of being a straight line as in most bivalves. One valve usually overhangs the other a little. This shell is perhaps most abundant on the east coast of Florida. It prefers a rocky or gravelly bottom, and a moderate depth of water.

ARCA CHEMNITZI Philippi   (Chemnitz's Ark)   p. 40
About an inch and a half in length, this species appears at first glance quite like the last. It is slightly inequivalve, short and high, and very well inflated. There are some twenty-five rounded or flattened ribs, crossed by distinct grooves. The beaks are more widely separated, and considerably higher, than with *incongrua*, and the basal line lacks the characteristic curve of the latter species. The color is white, and there is a sparse periostracum present in some of the radiating grooves. This bivalve occurs from Florida to Texas, and in the West Indies.

ARCA SECTICOSTATA Reeve   (Cut-ribbed Ark)   p. 40
This species ranges from the Carolinas to Texas. Its length is from two to three inches. The white shell is rather elongate and sturdy, being about twice as long as it is high. There is a brownish periostracum. The beaks are low, slightly incurved, and the hinge line is straight. There are about thirty-five radiating ribs which widen as they approach the basal margin, each rib bearing a deep central groove for most of its length. The teeth are small, but number more than fifty. This species is relatively common throughout most of its range. It lives in moderately shallow water, attached to rocks or corals by means of a strong, flat byssus, but it is able to move about and spin a new byssus at will. After a storm with an onshore wind one may find plenty of shaggy brown examples that have been torn from their moorings and cast ashore.

ARCA OCCIDENTALIS Philippi   (Turkey Wing)   p. 12
A common and familiar shell on southern beaches is the turkey wing, sometimes called the "Noah's Ark" shell. From two to four inches in length, the shell is sturdy, oblong, and gaping at both ends. The beaks are slightly incurved and rather widely separated, with a broad and flat area between them. The hinge is perfectly straight, with about fifty small teeth. This is a colorful clam, and fresh specimens are yellowish white, irregu-

larly streaked with reddish-brown bands. The interior is pale lavender. On living examples these colors are pretty well concealed by a thick and bristly periostracum. When one views a good specimen with both valves in place, it is easy to see why it has been dubbed "Noah's Ark." It used to be called *Arca noæ*, but that name is now restricted to an oriental species, so our clam has been renamed *occidentalis*. It may be found from Cape Hatteras to Cuba, living attached in rock crevices by means of a strong byssus. Most of the shells found on the shore are badly worn and bleached, but following storms one often finds fresh specimens, marked with strong ribs and tigerish streaks beneath the shaggy periostracum.

ARCA UMBONATA Lamarck   (Mossy Ark)   p. 40
The mossy ark is well named. The color of its shell is purplish white, inside and out, but the mollusk is almost completely covered with a dark brown, mossy periostracum, and frequently with growths of bryozoans and sponges. The shell is some two inches in length, boxlike, and elongate. The beaks are prominent and widely separated, with a broad area between them that is often scored with a geometric pattern. The anterior end of the shell is rounded, and the posterior end is sharply carinated. There are four or five strong ribs on the posterior end, and the rest of the surface is irregularly cross-ribbed by growth lines. The margins are smooth, not scalloped or crenulate as in many of the ark shells. The mossy ark may be found from North Carolina to the Gulf of Mexico. It lives in moderately deep water, but the shell is fairly common on beaches throughout its range. This clam occurs as a fossil in the Caloosahatchee beds (Pleistocene) south of Lake Okeechobee, Florida.

ARCA BARBATA Linne   (Bearded Ark)   p. 40
The bearded ark, or hairy ark, is another shaggy fellow, also occurring from the Carolinas to Texas. From one to two inches long and purplish brown in color, it differs appreciably from the last species by being broadly rounded at both ends, and in lacking the flattened area between the beaks. The valves are sculptured with numerous very fine, closely set, radiating ribs, each rib faintly beaded throughout its length. The hinge is typically comblike, but the teeth are small and few in number. As its name implies, the bivalve has a heavy, hairy periostracum. This species, too, prefers moderately deep water, but empty shells are generally to be found along rocky beaches on our southern shores.

ARCA CANDIDA Gmelin   (Bright Ark)   p. 40
This is a fairly large bivalve, at home from the Carolinas to Texas and the West Indies. Examples taken from the Atlantic Ocean are nearly three inches in length, while those collected from

the Gulf of Mexico are somewhat smaller. The shell is compressed, rounded at the anterior end, and bluntly pointed at the posterior end. The surface is sculptured with fine ribs and concentric growth lines, imparting a somewhat beaded appearance to the shell, most pronounced on the posterior slope. The hinge teeth are quite small. The color is white, with a soft and shaggy, yellowish-brown periostracum. Fresh specimens of this bivalve, now and then picked up on the beach after a heavy blow, are very bright and colorful objects with the golden-brown periostracum hanging over the edge of the shell like a fringe.

ARCA RETICULATA Gmelin   (Reticulate Ark)   p. 40
This species is only about one inch long, but it is strong and rugged, and moderately inflated. The posterior end is abruptly pointed, and there is a distinct ridge running from the posterior point to the neighborhood of the beaks. The shell has strong radiating ribs, and they cut across stronger concentric ridges, producing a distinctive network pattern. The hinge line is short, the teeth medium, and the margins of the valves are crenulate. The color is yellowish white, the periostracum yellowish brown. This decorative little bivalve lives in shallow water from Cape Hatteras south, commonly under sponges, corals, and other marine growths.

ARCA ADAMSI Smith   (Adam's Ark)   p. 40
This is a little fellow, only about one-half inch in length. The shape is oblong, with both ends rounded; the anterior one broadly and the posterior one more abruptly. The shell is relatively thick, and is decorated with numerous very fine beaded lines. The beaks are well elevated, and the hinge teeth are moderately large, but few in number. The margins are finely crenulate. The color is brownish olive or white. This species lives just offshore on rough bottoms, from North Carolina to the West Indian Islands.

Genus *Nœtia* Gray 1847

2 Species

NŒTIA PONDEROSA Say   (Ponderous Ark)   p. 40
The ponderous ark has a shell that is somewhat oblique, but quite thick and sturdy. The beaks are prominent and inclined to turn forward. The color is white, and there are some thirty strong ribs that are covered near the margins by deeply chiseled concentric lines. The comblike hinge teeth are larger at the ends, smaller at the center. In life this bivalve is covered with a heavy, velvety periostracum that is nearly black in color. When viewed end-on the two valves are remarkably heart-shaped. The ponderous ark lives in fairly deep water, commonly with marine growths attached to its shell. It occurs from Cape Cod

to the West Indies, rather uncommonly in the North, but very abundantly south of Cape Hatteras. Its average length is under three inches, but some individuals exceed that. There is no byssus.

Genus *Limopsis* Sassi 1827

11 Species

LIMOPSIS SULCATA Verrill & Bush (Sulcate Limopsis)
p. 192
This shell is about one-half inch in length, and its shape is obliquely oval. The posterior margin is produced and obtusely rounded, and the dorsal margin is short and straight. The beaks are small but prominent. There are about eighteen teeth, the posterior series curving downward. The valves are sculptured with rather strong concentric grooves, marked by faint vertical lines. The margins are crenulate. This shell is yellowish white in color, and is found in deep water in the Gulf of Mexico.

LIMOPSIS AURITA Brocchi (Eared Limopsis) p. 192
This species is about one-half inch in length, and, like the others of this genus, obliquely oval in shape. The valves are compressed, and covered with a hairy periostracum that is pale brown in color. Beneath this shaggy coat the shell is white. There is a sculpture of distinct concentric lines, especially marked on young shells. There are about five teeth in each series. This species occurs in the Gulf of Mexico, and has been dredged from a depth of thirty fathoms.

Genus *Glycymeris* Da Costa 1778

4 Species and varieties

GLYCYMERIS AMERICANUS Defrance (American Bittersweet)
p. 12
Members of this family, which used to be known as *Pectunculus*, are popularly called bittersweets. The American bittersweet is an inhabitant of moderately deep water and may be found from Cape Hatteras to the West Indies. About an inch and a half long in the northern parts of its range, it gets to be nearly three inches in the south. The shell is orbicular, well inflated, smooth and round, with the beaks centrally located. There is a curving row of comblike hinge teeth, becoming feeble or absent beneath the beaks. The surface bears numerous very fine striations, and the inner margins of the shell are crenulate. The color is creamy white, irregularly blotched with yellowish brown. Single valves are generally to be found along the southern beaches, but must be fairly fresh to show their true beauty.

GLYCYMERIS AMERICANUS LINEATA Reeve (Lined
Bittersweet) p. 192
This bivalve is more abundant in the West Indies, but examples
are occasionally taken as far north as North Carolina. It used
to be considered a separate species, but most authorities now
consider it as merely a variety of *americana*. It averages a little
more than one inch across, being somewhat smaller than the
typical form, and is very well inflated. The radiating lines are
more deeply incised, and the colors are about the same, with the
most colorful individuals coming from southern waters.

GLYCYMERIS PECTINATUS Gmelin (Comb Bittersweet)
p. 33
The comb bittersweet is a smaller species, about three-quarters
of an inch long, occurring from the Carolinas to Mexico. The
shell is small, rather compressed, with about twenty well-
rounded radiating ribs that are crossed by very minute growth
lines. There are about ten teeth on each side of the beak cavity.
The color is white, with spotted bands of yellowish brown. This
is a pretty little shell when obtained fresh and in good condition.
Specimens rolled about in the surf and left drying on the beach
lose much of their delicate color.

Family Pinnidæ

LARGE WEDGE-SHAPED bivalves, thin and fragile, and gaping at the
posterior end. They are attached by a large and powerful byssus.
Articles of wear, such as gloves, have been woven from the byssus
of a Mediterranean member of this family. There are large muscles
in both valves. Natives of warm seas, some grow to a length of
more than two feet, and occasional specimens contain black pearls.

Genus *Pinna* Linne 1758

2 Species

PINNA CARNEA Gmelin (Flesh Sea Pen) p. 192
This is a thin, wedge-shaped shell, more than a foot long when
fully grown, but most specimens found are less than that. The
posterior end is rounded and gaping, and the shell tapers to a
point at the other end, where the beaks are situated, so that one
might almost say there is no anterior end. Each valve is sulcate
longitudinally, and is decorated with a number of undulating,
radiating folds with small lines between them. The substance is
quite fragile, and the shells are easily broken. The color is
pale orange-yellow. The bivalve lives attached by a large and
silky byssus. The range is from Cape Hatteras to the West
Indies.

Genus *Atrina* Gray 1840

2 Species

ATRINA RIGIDA Dillwyn   (Stiff Sea Pen)   p. 41
This sea pen is a large but delicate shell, growing to a length of more than one foot. Olive-brown in color, the shell is thin, and triangular, or wedge-shaped, in outline. The dorsal margin is straight, the ventral margin rounded, and the posterior margin is gaping. The valves are not longitudinally sulcate, as in *Pinna*, but they are decorated with some fifteen rounded and slightly elevated ribs which fade near the beaks. Highly elevated tubular spines adorn the ribs, particularly near the outer margins. There is a very strong, silky byssus. This sea pen lives in fairly deep water, attached to some solid object. It occurs from North Carolina to South America, and shells, especially of smaller individuals, are often to be seen on the beaches, but the valves are so thin and fragile that they break up rather quickly.

ATRINA SERRATA Sowerby   (Saw-toothed Sea Pen)
                                                         p. 41
This is another sea pen, also living in a few fathoms of water from North Carolina to South America. This is a brown to brownish black bivalve, with the same general shape as the last species. It gets to be a little larger, and is much more delicately sculptured. The ribs are closely set, with scales that are smaller, but much more numerous, than with *rigida*. Large "scallops" offered for sale in southern markets are likely to be cut from the muscles of one of these sea pens.

Family Pteriidæ

IN THIS GROUP the shell is very inequivalve, the right valve with an opening under its wing for the passage of its byssus. Occurring in warm seas, this family contains the valuable pearl oysters.

Genus *Pteria* Scopoli 1777

4 Species

PTERIA COLUMBUS Bolten   (Winged Pearl Oyster)
                                                         p. 41
This bivalve has a moderately solid shell, three to five inches in length. The color is greenish brown with paler rays. The hinge line is straight, the posterior margin broadly rounded, and the wing is strongly notched. The surface is wrinkled, and young specimens are usually covered with prickly spines. Inside,

the shell is very pearly.　Formerly known as *Avicula atlantica* Lamarck, this is a member of the "pearl oyster" group.　The valuable shells of this group, however, are found in Ceylon and the Persian Gulf, although some pearl-fishing is done off Lower California and Panama.　This species occurs from North Carolina to the West Indies, usually in clusters attached to sea plants on the ocean floor.

Genus *Pedalion* Solander 1770

3 Species

PEDALION ALATA Gmelin　　　(Winged Tree Oyster)　p. 44
The winged tree oyster has a shell that is greatly compressed, the right valve practically flat and the left but little raised. The hinge line is short.　The surface may be smooth or scaly, and the color brown, black, or purplish, with juvenile specimens often rayed.　The inside of the shell has a pearly layer that does not extend all the way to the margins.　This species, which attains a length of three inches, thrives in shallow water along the Florida coast, generally living in massive colonies attached to the roots of mangroves or to submerged brush close to shore. In older books this shell may be found listed under the name of *Perna*.

PEDALION LISTERI Hanley　　　(Lister's Tree Oyster)

p. 192
This is an elongate shell, often somewhat irregular in outline. Its height is about an inch and a half, and its color greenish brown, commonly with paler rays.　The valves are compressed, the surface wrinkled, and the interior pearly.　There are about a half-dozen vertical grooves along the hinge line.　This bivalve is not overly common.　It may be found off the coast of southern Florida.

Genus *Pinctada* Roeding 1798

1 Species

PINCTADA RADIATA Leach　　　(Spiny Tree Oyster)　p. 41
In this species the shell is less oblique.　The valves are flatter and nearly equal in size.　There is a byssal notch under the right wing.　The surface is sculptured with scaly projections concentrically arranged, although these may be lacking in some specimens.　The color is variable, generally some shade of brown or green.　The shell is about two inches long.　Fresh specimens of this clam are often brought up by the sponge fishermen of Florida and empty valves are not uncommon on the beaches of that state. The interior of the shell is very pearly.

Family Ostreidæ

SHELL IRREGULAR and inequivalve, often large and heavy. The lower valve usually adheres to some solid object, and the upper valve is usually smaller. The distribution is world-wide, in temperate and warm seas. Probably the most valuable of the food-mollusks belong to this family.

Genus *Ostrea* Linne 1758

5 Species

OSTREA VIRGINICA Gmelin   (Virginia Oyster)   p. 44

The common oyster scarcely needs a description. From six to ten inches in length and lead-gray in color, the rough and heavy shell is generally narrow, elongate, gradually widening, and moderately curved, but it varies in surface and shape according to the position in which it lies during growth. The upper valve is smaller and flatter than the lower, and it moves forward as the shell advances in age, and growth of the ligament leaves a lengthening groove along the beak of the adhering valve. The interior is dull white, with the muscle-scar nearly central and deep violet.

This well-known shellfish is the most important commercial bivalve we have. The tiny youngsters (called "spat") are free-swimming for a short period before they settle upon some hard object to become sessile for life. Those individuals that chance to settle in the mud perish, and to minimize the annual loss oystermen spread tons of broken shells (called "clutch") over the beds each year. The oysters' chief enemies are the starfish and various snails, especially the oyster drill, *Urosalpinx cinerea* (page 77). The former wraps its five arms about the unlucky oyster and exerts a steady pull which may last for hours, until the bivalve's muscles are exhausted and it is forced to gape a little, and then the starfish is rewarded with a good meal. The little drill bores a neat round hole through the shell by means of its sandpaper-like tongue (radula), and feeds on the succulent parts within. The oyster is perhaps most commonly associated with Chesapeake Bay, but it thrives from Maine to Florida, and occurs locally as far north as Cape Breton Island. It has been successfully planted in California waters. Our oyster is almost identical with the European oyster, *Ostrea edulis*. Contrary to popular belief, valuable pearls are not found in the valves of this mollusk. The oyster's shelly lining is not pearly, but is smooth and dull, so any pearl that is formed is porcellaneous, and without luster or iridescence.

OSTREA FRONS Linne   ('Coon Oyster)   p. 44

The little 'coon oyster is a rosy brown to deep brown shell about one and one-half inches long, sometimes more. It is moderately thin and curved, and there is a broad longitudinal midrib, with coarse folds from midrib to margins. The attached valve bears several processes which clutch the stem of some sea plant or tree root. 'Coon oysters are found growing together in huge masses, often larger than a bushel basket. They occur throughout mangroves, and they range as far north as North Carolina. They derive their popular name from the fact that raccoons delight in feeding upon them.

OSTREA PERMOLLIS Sowerby   (Soft Oyster)   p. 44

Another popular name for this little fellow is the sponge oyster, because it lives, in the majority of cases, in masses of the "bread sponge." A little over an inch in length, it is a golden brown outside and bluish white inside. The shell is compressed, and very variable in outline, with a narrow hinge. The lower valve is flat, and the upper valve only slightly convex. The surface is wrinkled by irregular wavy ridges, and is covered with a periostracum that is thick but soft. This mollusk may be collected along the Florida coast, but it is not particularly abundant in most places.

Family Spondylidæ

SOMETIMES CALLED spiny oysters, these bivalves are attached to some object by their right valves. The surface is ribbed or spiny, and the hinge consists of two interlocking teeth in each valve. Many species are highly colored. They are confined to warm seas.

Genus *Spondylus* Linne 1758

2 Species

SPONDYLUS AMERICANUS Lamarck   (Spiny oyster)

p. 12

This shell has a couple of other popular names, thorny oyster and "chrysanthemum shell," the latter from its frilled, vari-hued appearance. It is a heavy and strong-shelled mollusk, from three to five inches long, and ranging from white to brown and purplish, and sometimes bright yellow or pinkish red. The shell is attached by its right valve, which has a broad, triangular hinge area, and crowded conditions often produce irregularly shaped individuals. The surface has many radiating ribs, and numerous scattered spines, some short and needle-like and some long and blunt. Interlocking teeth are present in each valve, and a pair can be opened only part way without fracturing them.

This colorful shell lives from the Carolinas south, but the largest and handsomest specimens come from the Gulf of Mexico. Shells are often found imbedded in chunks of coral that have been washed shoreward, and upper valves, badly worn, are not uncommon on the Florida beaches.

Genus *Plicatula* Lamarck 1801

2 Species

PLICATULA GIBBOSA Lamarck   (Cat's-Paw)   p. 44

The cat's-paw averages about one inch in length. The shell is small but solid, and fan-shaped, with six or seven broad folds radiating from the beaks. The right valve is the larger of the two, and the mollusk is attached to some firm object by the umbo of this valve. The color is white, often with gray or reddish lines. One may find plenty of these little fan-shaped shells in the drift alongshore, all the way from Cape Hatteras to Florida, but the delicate, pencil-like coloring fades rapidly when the shell is exposed to the sun's rays, and most of the specimens picked up will be a lusterless white. A larger species is sometimes collected in Florida, occasionally as much as two inches in length. This is *Plicatula spondyloidea* Meuschen. This form often grows in crowded clusters, one individual attached to another. The hinge teeth are sturdy, and the valves are not easily separated. This bivalve ranges south to Texas, but is not very abundant.

Family Pectinidæ

THESE ARE the scallops. The valves are commonly inequal, the lower one strongly convex, the upper one flat or even concave. The surface is usually ribbed, and the margins scalloped. Juvenile specimens are attached by a byssus, while adults are generally free-swimming. There is a row of tiny eyes fringing the edge of the mantle, each complete with cornea, lens, and optic nerve.

Scallops are found in all seas, from shallow water to great depths. Shells of this family have always had a certain artistic appeal. The Crusaders used the shell of *Pecten jacobæus* from the Mediterranean as a badge of honor, and at present a scallop shell is the well-known trade-mark of a great oil company.

Genus *Pecten* Osbeck 1765

48 Species and varieties

PECTEN IRRADIANS Lamarck   (Bay Scallop) pp. 13, 45

This is a medium-sized bivalve, attaining a length of about three inches. The color is variable, ranging from nearly white through orange, reddish, purplish, and mottled brownish. Often there are concentric bands. The shell is roughly round, the valves convex, the lower one less so than the upper. There are about twenty elevated, rounded ribs, the spaces between them similarly rounded, their surface marked with many fine, concentric growth lines. The ears (wings) are nearly equal in size, and are covered with small, radiating ridges. The interior is purplish white, the muscle-scar small and shallow. Unlike most of the clams, the scallop does not crawl or burrow. It progresses through the water by a series of jerks and darts produced by rapidly opening and shutting its valves. This manner of locomotion is powered by a large, single muscle, which is the part of the scallop we eat. This group is also unusual among pelecypods in that its members possess functional eyes, and a resting scallop may be approached by a crab or a carnivorous snail, but before the predator gets too close for comfort the bivalve literally "leaps" upward and goes skittering off to safety.

This is the common scallop of the Atlantic coast from New England to Cape Hatteras, and tons are dredged annually for the markets. It occurs in shallow water, commonly living amongst the eel-grass, and the unfortunate disappearance of the eel-grass in many localities has led to a corresponding scarcity of scallops. Fishermen obtain them by dragging a rakelike instrument through this grass.

PECTEN GIBBUS Linne (Calico Scallop) pp. 13, 45
The little calico scallop occurring from North Carolina to Cuba is one of the commonest of the bright-colored shells to be found on our southern beaches, and thousands of them are used yearly in making various shell novelties. From one to one and one-half inches in length, the shell is inflated, the lower valve more than the upper. There are about twenty strong radiating ribs, marked by numerous growth lines, giving the shell a somewhat rough surface (when not wave-worn). The wings are about equal in size. The color patterns exhibited by this little clam are very numerous; various combinations of mottled white, rose, brown, purple, and orange-yellow, with the colors more striking on the upper valve. A variety of this scallop, *amplicostatus* Dall, gets to be some two inches in length. The dorsal valve is mottled gray and black, and the ventral valve is usually pure white, sometimes tinged with yellow on the umbo. This variety ranges from Florida to Texas, and south to South America. An example is shown on page 45.

PECTEN MUSCOSUS Wood (Rough Scallop) p. 13
The rough scallop occurs from the Carolinas to the West Indies. About two inches long, it is a shallow water species, often seen in tide pools. Pinkish red to deep reddish brown in color, some individuals are bright lemon yellow. The shell is small but sturdy, decorated with about twenty strong ribs, each composed

of a bundle of smaller ribs. The lower portions of these ribs are studded with erect, sharp scales, giving the shell a rough, spiny appearance. The wings are unequal in size. This species used to be known as *Pecten esasperatus* Sowerby.

PECTEN RAVENELI Dall   (Ravenel's Scallop)   p. 45
This is a pinkish to purplish, and occasionally golden yellow, scallop about two inches across, occurring in Florida. The upper valve is quite flat, and deeply colored with irregular dark markings. The lower valve is very concave, and decorated with about twenty strong, grooved ribs with wide spaces between them. The hinge line is straight, the wings unequal, and the basal margin forms an almost perfect semi-circle. This bivalve, formerly called *Pecten hemicyclica*, is sometimes confused with the next, *ziczac*, but it is a smaller shell, more lightly colored, and it has a different sculpture on the lower valve.

PECTEN ZICZAC Linne   (Sharp-turn Scallop)   p. 60
This larger scallop has a flat upper valve, often with a central concavity, and a lower valve that is deep and cup-shaped, and overlapping the upper. The lower valve has low radiating ribs, so low that the surface seems rather smooth. The flat upper valve is heavily mottled, and ornamented with zigzag lines of black. The lower valve is mottled brown, verging on reddish. This is a medium-sized scallop, about three inches across, that is abundant in Bermuda and taken occasionally during the summer months in Florida.

PECTEN SENTIS Reeve   (Thorny Scallop)   p. 13
This is a small species, averaging little more than one inch in length. It is often bright scarlet in color, but may be orange-brown, purple, or even white, usually with some shading. Both valves are generally colored the same. The wings are very unequal. In many ways this species looks like the next one, *ornatus*, but the thorny scallop has more numerous ribs, about forty, each with tiny but sharp scales. This is generally a common species on Florida beaches, mostly in the form of single valves.

PECTEN ORNATUS Lamarck   (Ornate Scallop) pp. 13, 45
This is a fragile little bivalve living from southern Florida to the West Indies. About one inch long, it is white, spotted with red and purple. There are about twenty strong ribs, a few of which are usually unspotted. The ribs are studded with short but sharp spines, especially near the margins. The wings are very unequal in size, one of them being scarcely noticeable. Like the other small varieties so far discussed, this species may be seen in shallow water, darting about among the seaweeds. Beach-worn shells have usually lost their scales or spines.

PECTEN NODOSUS Linne   (Lion's-Paw)   p. 13

The lion's-paw is a handsome shell, eagerly sought by collectors. From four to six inches across, the shell is heavy and robust, with valves about equal in size and shape. They are but little arched. The surface has numerous closely spaced radiating ribs, and about ten broad folds, the crests of which are marked at regular intervals with blunt, raised knobs. The interior has channels corresponding to the outside folds. The wings are not equal in size. The color is reddish brown to bright orange, with the glossy interior generally some shade of pink or salmon. The lion's-paw lives in rather deep water, from Cape Hatteras to the Florida Gulf coast, but it is seldom found on the beach. Single valves can be picked up at Sanibel Island after storms, but most of the good specimens seen are brought in by sponge fishermen.

PECTEN GRANDIS Solander   (Deep-sea Scallop)   p. 45

The deep-sea scallop, or giant scallop, has a shell that is large, orbicular, somewhat higher than long, and moderately thick and solid. It commonly gets to be six or seven inches across. The lower valve is nearly flat, the upper slightly convex. The surface is sculptured with a multitude of narrow radiating lines and grooves, and the wings are about equal in size. The upper valve is reddish brown, rarely rayed with white, and the lower valve is pinkish white. Inside, the shell is glossy white, with a very prominent muscle-scar. This is the largest American scallop. The convex valves were commonly used as dishes by the Indians, and today visitors to our northern shores nearly always take home a shell or two to be used as ash trays. This is a deep-water form, occurring from Newfoundland to New Jersey. In older books this species may be found listed as *Pecten magellanicus* Gmelin, or *Pecten tenuicostatus* Mighels.

PECTEN ISLANDICUS Muller   (Iceland Scallop)   p. 60

This is another northern bivalve, growing to a length of some four inches. The shell is oval, the upper valve more convex than the lower. Pale orange to reddish brown in color, the surface has about fifty narrow, unequal, crowded, radiating ridges bearing numerous small, erect scales. These ridges are frequently grouped so as to form a number of unequal ribs. The wings are very unequal, the posterior one being shorter. The interior is glossy white, the muscle-scar large and shallow. This species is a lover of cold water, and is found from Iceland to Maine, as well as in northern Europe.

PECTEN IMBRICATUS Gmelin   (Little Knobby Scallop)
     p. 192

This distinctive little scallop is unusually flat, the valves scarcely arched at all. Its length is about an inch and a half, and its height a little more, the shape being somewhat triangular. There are nine or ten stout ribs, each with a series of regularly spaced hollow knobs. The wings are very unequal in size. The color is mainly white, sometimes variegated with pinkish, and the inside of the shell is yellowish, with the margin and the hinge area purplish. This species is quite uncommon, living off the coast in southern Florida.

PECTEN VITREUS Gmelin   (Transparent Scallop) p. 192
This is a nearly round little shell, about one inch across, frequently smaller. The valves are very thin and translucent, and but little inflated. The wings are unequal. The surface appears to be smooth and glossy, but under a lens it shows numerous very small radiating lines, and several rows of tiny beads arranged concentrically near the margins. The color is silvery gray. This is a deep-water form, occurring from Massachusetts to Labrador.

PECTEN STRIATUS Muller   (Striate Scallop)   p. 192
This is a small and rather delicate scallop, about three-fourths of an inch in length. The valves are quite round, thin, and moderately inflated, the wings slightly unequal. The surface looks smooth, but is really decorated with numerous very fine radiating striæ. The color is white, variously marked with pinkish or rose. This little bivalve lives in cold water, at considerable depths. It has been taken off Martha's Vineyard at 100 fathoms, and it occurs as well in Scotland and northern Europe.

PECTEN DALLI Smith   (Dall's Scallop)   p. 60
This species, listed for years as *Amusium dalli*, is an odd little scallop, something more than an inch across. It is an extremely thin and fragile shell, orbicular in outline, and considerably compressed. The outside is smooth, with several low but distinct concentric ribs near the margin, while the interior bears a number of prominent radiating ribs. The wings are small, and the valves gape a little at each end. This is a rather deep-water pelecypod, occurring in the Gulf of Mexico and in the West Indies.

Family Limidæ

THESE SHELLS are obliquely oval, and usually winged on one side. The ends are gaping. The hinge is toothless, with a triangular pit for the ligament. The color is generally white. Members of this family are often called file shells. They are as expert in swimming as the scallops, but they dart about in the opposite direction, with

MISCELLANEOUS PELECYPODS

All shells approximately one-half natural size.

1. *Asaphis deflorata* 1 view: **RAYED COCKLE**   p. 83
   Fine radiating lines; purplish.

2. *Loripinus chrysostoma* 1 view: **BUTTERCUP**   p. 58
   Orbicular; yellow or orange inside.

3. *Papyridea hiatus* 1 view: **SPINY PAPER COCKLE**   p. 64
   Thin-shelled; pinkish; fine spines.

4. *Tellina alternata* 1 view: **LINED TELLIN**   p. 75
   Yellowish white; sharp concentric lines; polished.

5. *Tellina interrupta* 1 view: **INTERRUPTED TELLIN**   p. 74
   Elongate; purplish radiating marks; not polished.

6. *Tellina radiata* 1 view: **RISING SUN**   p. 74
   Elongate; rosy radiations; highly polished.

7. *Cardita floridana* 1 view: **BIRD SHELL**   p. 54
   Small but solid; broadly rounded ribs.

8. *Tellina lineata* 2 views: **ROSE PETAL**   p. 74
   Rosy pink, inside and out; not polished highly.

9. *Chione cancellata* 1 view: **CROSS–BARRED VENUS**   p. 69
   Sturdy; network pattern; sometimes banded.

10. *Chione interpurpurea* 1 view: **MOTTLED VENUS**   p. 69
    Concentric grooves; many sharp scales.

11. *Pitar dione* 1 view: **ELEGANT VENUS**   p. 68
    Pale lilac; usually with long spines.

12. *Chione latilirata* 1 view: **BROAD–RIBBED VENUS**   p. 69
    Broadly rounded concentric folds.

13. *Macrocallista maculata* 1 view: **CHECKERBOARD**   p. 67
    Rounded; heavily blotched; polished.

14. *Strigilla carnaria* 2 views: **ROSY CARNARIA**   p. 79
    Small; roundish; rosy pink.

15. *Macrocallista nimbosa* 1 view: **SUN–RAY SHELL**   p. 67
    Elongate; violet gray, finely rayed; polished.

Plate 4   29

MISCELLANEOUS GASTROPODS

All shells approximately one-half natural size.

1. *Nerita versicolor* 2 views: **VARIEGATED NERITE**   p. 118
   Black, white, and red.

2. *Purperita pupa* 3 views: **ZEBRA SHELL**   p. 119
   Zebra-like pattern.

3. *Nerita tessellata* 2 views: **CHECKERED NERITE**   p. 119
   Black and white — no red.

4. *Neritina reclivata* 1 view: **GREEN NERITE**   p. 120
   Dark greenish; white inner lip.

5. *Neritina virginea* 3 views: **VIRGINIA NERITE**   p. 119
   Polished; innumerable color combinations.

6. *Cerithium literatum* 1 view: **LETTERED HORN SHELL**
                                                          p. 159
   Elongate; marked with scrawls suggesting letters.

7. *Nerita peloronta* 2 views: **BLEEDING TOOTH**   p. 118
   Orange stain around white tooth.

8. *Olivella jaspidea* 2 views   p. 214
   Highly polished; no plaits on columella.

9. *Cypræa spurca* 1 view: **LITTLE YELLOW COWRY**   p. 170
   Small; yellowish, spotted with white.

10. *Cyphoma gibbosa* 1 view: **FLAMINGO TONGUE**   p. 167
    White and polished; "humpbacked."

11. *Natica canrena* 1 view: **SPOTTED MOON SHELL**   p. 132
    Streaked with yellow and brown; shelly operculum.

12. *Astræa longispina* 1 view: **STAR SHELL**   p. 116
    Flatly coiled; pearly; long spines.

13. *Cypræa exanthema* 1 view: **MEASLED COWRY**   p. 167
    Large; polished; heavily spotted.

14. *Cypræcassis testiculus* 1 view: **BABY BONNET**   p. 174
    Fine longitudinal grooves; reflected inner lip.

15. *Turritella exoleta* 1 view: **TURRET SHELL**   p. 150
    Spike-like; angled whorls.

16. *Oliva sayana* 2 views: **LETTERED OLIVE**   p. 213
    Elongate; highly polished; plicate columella.

17. *Pisania pusio* 1 view: **PISA SHELL**   p. 197
    Fusiform; strongly spotted.

18. *Astræa cælata* 1 view: **CARVED STAR**   p. 117
    Rugged and strong; knobby surface.

19. *Haminæa elegans* 1 view: **GLASSY BUBBLE**   p. 226
    Thin and fragile; glassy.

20. *Livona pica* 1 view: **MAGPIE SHELL**   p. 112
    Strong blacks and whites; pearly under surface.

the hinge foremost, often trailing a long sheaf of filaments. Some of them build nests of broken shells and coral, and sand grains, held together by byssal threads. There are many fossil forms.

Genus *Lima* Bolten 1798

6 Species

LIMA LIMA Linne   (File Shell)   p. 60
The common file shell is indeed well named, as its rasplike surface does very strongly remind one of a file. It is about one and one-half inches long, sometimes longer, with a white, moderately thick, oblique shell, the valves only slightly gaping. The surface bears about twenty broad radiating ribs, each with many closely set erect scales. It lives in fairly shallow water, and enjoys a wide distribution, being found in the Mediterranean Sea as well as in southern Florida. It used to be called *Lima squamosa* Lamarck.

LIMA SCABRA Born   (Rough File Shell)   p. 13
The rough file shell is a fine example of this group of free-swimming clams, attaining a length of nearly four inches. It is a moderately thick and heavy shell, oval in shape, and not as oblique as most of the group. The valves are rather compressed, gaping a little at the hinge, and decorated on the surface with closely set ridges covered with small pointed scales. It usually bears a thin, yellowish-brown periostracum. This species occurs in southern Florida and the West Indies.

LIMA INFLATA Lamarck   (Inflated File Shell)   p. 60
This shell is about one and one-half inches long, and pure white in color, with the animal itself bright orange. The shell is quite thin but sturdy, oblique, considerably inflated, and gaping at both ends, so that the valves are in contact only at the hinge and the basal margin. The surface is sculptured with fine ribs, often with riblets between them. The inflated file is to be found from Cape Hatteras to Trinidad. It is a shallow-water form and lives in crevices and under stones in bays and lagoons. It attaches itself by means of a byssus, and frequently builds a crude sort of nest of byssal threads, plastered with bits of seaweed and pebbles, but the mollusk can cast all this aside and go zigzagging off with speed and dispatch when the occasion demands action.

LIMA TENERA Sowerby   (Delicate File Shell)   p. 60
This species is a little more than one inch long as a rule, with some individuals larger. It is pure white in color, and oval in outline, gaping a little at the posterior end, and more so at the anterior end. The surface is roughened by fine radiating lines

which are notched by small, sharp scales, so that the shell has a satiny luster. This file shell is rather easily recognized by its delicate sculpture. Like the others in this group it lives among the crevices to be found on rocky bottoms, and swims freely about at will. It prefers deeper water, and may be collected on both the east and west coasts of Florida.

LIMA HIANS Gmelin (Gaping File) p. 192
This is a rather small form, generally less than two inches high. It has the same oblique shape as the other file shells, is somewhat compressed, and gapes widely at the anterior end. The surface is roughened by numerous small but sharp radiating lines, especially prominent at the center of the shell. The color is white. This species occurs in southern Florida and the West Indies, and is also found in Europe.

Genus *Limatula* Wood 1839

7 Species

LIMATULA HYALINA Verrill & Bush (Little File)
p. 192
The name "little file" could apply equally well to almost any of this group. This species is a tiny fellow, about a quarter of an inch long. The shell is white, thin and translucent, and obliquely oval in shape, the posterior end extended. The hinge is short and straight. The surface bears some twenty sharp radiating ridges, separated by wider concave intervals. This bivalve lives in moderately deep water, off the Florida coast.

Family Anomiidæ

THESE ARE thin, translucent clams, usually pearly inside. They are attached to some solid object by a stalklike byssus which passes through an opening in the lower valve. The byssus becomes calcified, and the bivalve is permanently fixed. They are natives of warm and temperate seas.

Genus *Anomia* Linne 1758

2 Species

ANOMIA SIMPLEX Orbigny (Jingle Shell) p. 44
This shell is about one inch long and varies considerably in color, ranging from sulphur-yellow to coppery red, with many specimens silvery gray or black. The shell is circular, and variously distorted according to the object to which it is attached. The margins are sometimes undulating or jagged. The surface is minutely scaly and of a waxy luster. The upper valve is very

VARIOUS MOLLUSKS

1. Internal shell (pen) of the *Sepia officinalis* (Mediterranean) × 1, 2 views: **CUTTLEFISH**   p. xix

2. Internal shell of *Spirula spirula* × 1, 3 views, one showing the septa: **RAM'S HORN**   p. xix

3. Scaphapod *Dentalium semistriolatum* × 1, 1 view: **TUSK SHELL**   p. xix

4. *Argonauta americana* × 1, 2 views: **PAPER ARGONAUT**
                                                              p. xix

5. *Chiton tuberculatus* × 1, 1 view: **CHITON**   p. xix

6. *Nautilus pompilius* (Polynesia) × 1, 1 view: **PEARLY NAUTILUS**   p. xix

Plate 6   33

THE NUT SHELLS

1. *Nucula proxima* × 2 and × 5, 5 views: **NEAR NUT SHELL**
   p. 6
   Pearly interior, crenulate margin.

2. *Nuculana acuta* × 2, 3 views: **POINTED NUT SHELL** p. 7
   Acutely pointed; interior margin non-crenulate.

3. *Nuculana tenuisulcata* × 2, 3 views: **SULCATE NUT**   p. 7
   With strong concentric ribs.

4. *Solemya velum* × 2, 2 views: **AWNING SHELL**   p. 5
   Ragged, fringelike covering.

5. *Yoldia sapotilla* × 1, 2 views   p. 8
   Smooth, shining, greenish.

6. *Glycymeris pectinatus* × 1, 2 views: **COMB BITTERSWEET**
   p. 18
   Interior with curving tooth-series.

7. *Yoldia thraciæformis* × 1, 2 views: **BROAD YOLDIA**   p. 9
   Broad, pit between tooth-series.

8. *Arca pexata* × 1, 1 view: **BLOODY CLAM**   p. 11
   Strongly ribbed; animal has red blood.

9. Hinge line of typical *Arca*

convex, with a small beak, while the lower valve is smaller, flat, and has a circular hole for the passage of a fleshy byssus by which the mollusk adheres. These are the "jingle shells," greatly prized by children at the seashore, and perhaps the most abundant and familiar shell on many beaches. The bivalve lives attached to stones or other shells, out in moderately deep water. The cuplike upper valve is the one that is washed ashore after the animal dies, and the perforated lower valve is not very often seen. This species is found along the whole Atlantic coast.

ANOMIA ACULEATA Muller   (Prickly Jingle Shell)

p. 44

The prickly jingle shell is only about one-half inch in length, and it is yellowish white in color. The shell is small and rounded, the upper valve convex and the lower one thin and flat. The surface of the upper valve is covered with minute, prickly scales, commonly arranged in radiating rows. The interior is purplish white. The prickly jingle shell occurs with the larger *simplex*, all along the Atlantic coast. It is easily distinguished by its smaller size and prickly upper surface. It lives attached to stones and seaweeds, and clusters of specimens are often found on the roots of sea plants cast up on the beach after a strong inshore blow.

Family Mytilidæ

SHELLS EQUIVALVE, the hinge line very long and the umbos sharp. Some members burrow in soft mud, clay, or wood, but the majority are fastened by a byssus. These are the mussels, most of which are edible, though seldom eaten in this country. In Europe the mussel is "farmed" much the same as the oyster is here, and it forms an important item of European sea food. Mussels are world-wide in distribution, being best represented in cool seas.

Genus *Mytilus* Linne 1758

4 Species and varieties

MYTILUS EDULIS Linne   (Blue Mussel)   p. 61

The familiar blue mussel is about three inches long and bluish violet to blue-black in color. The shell is roughly an elongate triangle, the beaks forming the apex. The anterior margin is generally straight, the posterior margin broadly rounded. The surface bears many fine concentric lines and is covered with a shining periostracum. The interior is white, the margins violet. Young specimens are usually brighter colored, and may be greenish or even rayed. Acres of blue mussels are exposed at low tide all along the eastern seaboard. We find them attached to the stones and pebbles where the water is clear, and on the pilings of wharves, and in rocky places generally. They are attached by a series of strong byssal threads, but are capable of moving about to some extent. This clam is a tasty morsel, much relished by those who have tried it, but for some reason, probably the abundance of larger bivalves, it has never become very popular here, although this same species is eaten by the ton in Europe.

MYTILUS RECURVUS Rafinesque   (Bent Mussel)

p. 12

The bent mussel is greenish brown to purplish black, and it gets to be about one and one-half inches long. The shell is moderately solid, triangular, slightly inflated, and strongly and obliquely curved. The surface is decorated with a pattern of fine, elevated lines which often divide as they approach the posterior end. The interior is polished, and purplish in color, with a whitish margin. This little mussel, formerly known as *Mytilus hammatus* Say, is abundant in Florida waters, and may be found as far north as New Jersey. Clusters of individuals attach themselves to the roots of mangroves, to rocks and dead shells in shallow water, and even to living oyster shells.

MYTILUS EXUSTUS Linne   (Scorched Mussel)   p. 61

This species is about one inch long and bluish gray in color, with a bright brownish periostracum. The shell is small, thin, and somewhat fan-shaped on one side, giving the outline a peculiarly one-sided appearance. The surface is ribbed, strongest near the margins. The scorched mussel is a common species, generally washed ashore in clusters attached to other shells. It may be recognized by its lopsided outline. It is found from Cape Hatteras to Mexico and the West Indies.

Genus *Volsella* Scopoli 1777

11 Species and varieties

VOLSELLA MODIOLUS Linne   (Horse Mussel)   p. 61

The horse mussel is a big fellow, five or six inches long. The shell is heavy and coarse, and oval-oblong in shape. The beaks are placed slightly to one side. The anterior end is short and narrow, while the posterior end is broadly rounded. The surface is marked by lines of growth, and sometimes by a few faint radiating lines. The color is bluish black, with a thick and leathery periostracum. The interior is white. The horse mussel inhabits deeper water, dwelling on a rocky bottom where it is able to find secure places of attachment. The members of this genus commonly spin a nest, using pebbles and shell fragments combined with byssal threads to construct a safe refuge on the ocean floor. This large mussel, generally considered unfit for food, occurs from Arctic seas to the region of Cape Hatteras. The empty valves, noticeable on account of their size, are thrown up on almost every beach that is exposed to the open sea. Bleached shells often turn a pale lavender or reddish.

VOLSELLA TULIPUS Linne (Tulip Mussel) p. 61
The tulip mussel, from two to three inches in length, is yellowish brown in color, with the interior dark purplish. The mussel is shaped much like the last species, the anterior end short and narrow and the posterior end broadly rounded, but the shell is much thinner and moderately inflated. The beaks are anterior, not terminal. The surface is smooth, with a glossy periostracum. The tulip mussel is a pleasingly formed bivalve, found rather commonly all along our southern coasts.

VOLSELLA PLICATULUS Lamarck (Ribbed Mussel)
p. 61
The ribbed mussel, sometimes listed as *Volsella*, or *Modiolus dimissus* Dillwyn, is some three inches in length and yellowish green to bluish green in color. The shell is moderately thin, oblong-oval, and much elongated. The beaks are prominent, placed slightly to one side, but nearly in contact with the anterior extremity, which is small and rounded. The surface is ornamented with numerous radiating, somewhat undulating ribs that occasionally branch. The interior is silvery white, often iridescent. These are the familiar mussels of brackish water and tidal flats. Here they are found clustered in among the stones or imbedded in the peatlike earth, near the high water mark, their valves frequently encrusted with bryozoans and barnacles. This species is not generally regarded as edible, as it appears to thrive best in partially polluted waters. Its range is from Nova Scotia to Florida.

VOLSELLA ABORESCENS Dillwyn (Paper Mussel)
p. 192
This is a very thin and delicate little shell, just about one inch long. The shell is elongate, almost cylindrical, with low beaks situated at the anterior end, but not terminal. The color is greenish white, often iridescent. This thin and fragile mussel occurs from Florida to Texas and in the West Indian Islands. It lives in the muds between the tides.

VOLSELLA OPIFEX Say (Artist's Mussel) p. 193
This is a tiny mussel, usually less than one-half inch long, generally found attached to some other shell, and partially encased in a small round mound of cemented sand grains. The color is reddish brown, with the interior highly iridescent. The outline is somewhat cylindrical, the tips drawn out to a point and often frayed, so that the appearance is quite like the pointed end of a

small brush. As a result, this species is popularly known as the "artist's mussel." Thomas Say, the discoverer of this pelecypod, wrote as follows, in 1826, ". . . on a single valve of *Pecten nodosus* were several elevations that, on a cursury glance, presented an appearance not unlike the *Balanus* (barnacle). On a more particular inspection, each elevation prooved simular to the others . . . and composed of fine dark sand agglutinated together, attached by a broad base to the surface of the *Pecten*, and rising in the shape of a very low cone around an included shell . . . with its byssus very firmly affixed to the supporting surface."

Genus *Botula* Morch 1853

2 Species

BOTULA FUSCA Gmelin   (Dusky Mussel)   p. 193
This is an oddly shaped shell, also about one-half inch in length. The beaks are situated at one end, which they sometimes overhang, producing a hooked effect. The shell is rather cylindrical and somewhat curved. The surface is smooth, but coarsely wrinkled by growth lines. The color is dark chestnut-brown, with a shining periostracum. This species ranges from North Carolina to the West Indies.

Genus *Dacrydium* Torrell 1859

1 Species

DACRYDIUM VITREUM Moller   p. 193
This tiny bivalve is only about one-eighth of an inch long, and as might be suspected, it has no common or popular name. Its shape is somewhat triangular, with a short and acutely rounded anterior end, and a deep and broadly rounded posterior end. The beaks are low. The valves are thin, considerably inflated, and the color is a vitreous white, clean and shining. This is a deep-water pelecypod, occurring from the Arctic Ocean to off Florida, and it has also been taken in the English Channel, the Azores, and the Mediterranean.

Genus *Idas* Jeffreys 1876

1 Species

IDAS ARGENTEUS Jeffreys   p. 200
This is an oblong shell, less than one-quarter of an inch in length, living in deep water off the coasts of both Europe and North America. It has been dredged from 335 fathoms south of Martha's Vineyard, Massachusetts. The shape has been described as like a miniature *Arca*, but the hinge is toothless. The surface is smooth, and the color is pale brown. This is another mollusk that is too uncommon to have acquired a popular name.

Genus *Lithophaga* Bolten 1798

4 Species

LITHOPHAGA ANTILLARUM Orbigny   (Antillarian Date)   p. 61

The date shell attains a length of about two inches. The shell is thin, elongate, cylindrical, and wedge-shaped when viewed from above. The beaks are low and insignificant, and the hinge line is without teeth. The surface bears numerous concentric furrows, most pronounced near the posterior end. The color is brown, and the thin periostracum is also brown. The date shells, which do bear a striking resemblance to the seeds of dates, occur in southern waters. When young they suspend themselves to rocks by a byssus, but when adult they form cavities corresponding to the shape of their valves in limestone and other moderately soft rocks.

LITHOPHAGA NIGRA Orbigny   (Black Date)   p. 61

The black date shell is a smaller species, averaging a little more than one inch in length. Dark brown to nearly black in color, the shell is cylindrical, much like the larger date shell just described, with the periostracum somewhat thicker. The surface has strong concentric growth lines, crossed near the base by prominent vertical striations. This is a burrowing bivalve, excavating cavities in limestone, coral, and other semi-hard substances. The fleshy parts of the animal are luminous in the dark. This shell occurs from South Carolina to South America.

LITHOPHAGA BISULCATA Orbigny   (Two-furrowed Date)

p. 193

About one inch in length, this is a smooth and somewhat polished shell bearing a pair of weak furrows radiating from the beaks. The anterior end is bluntly rounded, and the longer posterior end is abruptly tapering. The color is pale brown, but the surface is generally covered by a calcareous incrustation. This boring clam occurs from North Carolina to the West Indies.

Genus *Modiolaria* Beck 1838

5 Species

MODIOLARIA NIGRA Gray   (Little Black Mussel)

p. 61

This is a small, brownish black mussel, averaging about one inch in length. The shell is thin and oval, and slightly produced at the posterior end. The beaks are fairly prominent, and placed some little distance from the anterior end. The surface bears a network of very minute, crowded lines and numerous fine radiating lines, the latter lacking on an area midway of the shell. The periostracum is rusty brown, and the interior of the shell is pearly. This species is generally not as abundant as the larger mussels, but they may be found in rock crevices and attached to empty shells from Cape Hatteras northward. It is more active than most of the others, and easily moves from place to place, using its foot as a prehensile organ, and spinning a new byssus when a satisfactory situation has been found.

Genus *Crenella* Brown 1827

6 Species

CRENELLA DECUSSATA Montagu   (Little Round Mussel)   p. 61
This is a small, brownish, well-rounded mussel, about one-half inch in length. The shell is obliquely oval, inflated, and rather fragile. The surface is sculptured with numerous crowded radiating lines, often appearing beaded owing to faint concentric lines. The inner surface is pearly and the margins are crenulate. There is a thin brownish periostracum. This bivalve lives in the mud just offshore from Greenland to North Carolina, and it is a rather common shell in the stomachs of bottom-feeding fishes.

CRENELLA FABA Muller   (Little Bean Mussel)   p. 193
This is a tiny fellow, not more than one-fourth inch long. Its shape is oval, with a small portion of the hinge line straight. The beak end is acutely rounded, while the other end is broadly rounded. The surface is sculptured with numerous distinct radiating lines which make the margins crenulate. The color is yellowish brown, with the interior lead-color, but polished. This little mussel lives in the mud from moderately shallow to deep water, from Maine north.

Family Dreisseniidæ

SMALL MUSSELS with a shelflike plate under the beaks for the attachment of the anterior muscle. Found in shallow water.

Genus *Congeria* Partsch 1835

3 Species

CONGERIA LEUCOPHEATA Conrad   (Platform Mussel)
p. 192
This is an elongate, partially curved little mussel, about one inch long. The valves are moderately well inflated, and bluntly keeled. The beaks are terminal, and the beak-cavity contains

THE ARK SHELLS

1. *Arca reticulata* × 2, 2 views: **RETICULATE ARK**   p. 16
   Network sculpture.

2. *Arca auriculata* × 1, 1 view: **EARED ARK**   p. 11
   Large and sturdy; back end square.

3. *Nucula verrillii* × 2, 2 views: **VERRILL'S NUT SHELL** p. 6
   Tiny; triangular; brownish.

4. *Arca adamsi* × 2, 2 views: **ADAM'S ARK**   p. 16
   Small; ribs crowded.

5. *Nœtia ponderosa* × ⅔, 2 views: **PONDEROUS ARK**   p. 16
   Heavy and solid.

6. *Nucula tenuis* × 2, 2 views: **THIN NUT SHELL**   p. 6
   Small; oval; greenish.

7. *Arca secticostata* × ⅔, 2 views: **CUT–RIBBED ARK**   p. 14
   Ribs longitudinally grooved.

8. *Arca barbata* × 1, 2 views: **BEARDED ARK**   p. 15
   Hairy in life.

9. *Arca candida* × ⅔, 2 views: **BRIGHT ARK**   p. 15
   Not much inflated; ribs fine.

10. *Tindaria amabilis* × 2, 2 views   p. 10
    Small; roundish; teeth prominent.

11. *Arca incongrua* × 1, 1 view: **INCONGRUOUS ARK**   p. 11
    Ribs cross-checked; basal line curved.

12. *Arca transversa* × 1, 2 views: **TRANSVERSE ARK**   p. 11
    Oblong; beaks low.

13. *Arca chemnitzi* × 1, 1 view: **CHEMNITZ'S ARK**   p. 14
    Ribs cross-checked; basal line straight.

14. *Arca umbonata* × 1, 3 views: **MOSSY ARK**   p. 15
    Purplish; rough and shaggy.

Plate 8   4I

THE SEA PENS AND OTHERS

1. *Pteria colymbus* × I, I view: **WINGED PEARL OYSTER**
   p. I9
   Winged, pearly inside.

2. *Chama sarda* × I, 2 views: **LITTLE JEWEL BOX**   p. 55
   Small but solid; thick-shelled.

3. *Pinctada radiata* × I, 2 views: **SPINY TREE OYSTER**   p. 2c
   Commonly frondose.

4. *Atrina rigida* × ½, I view: **STIFF SEA PEN**   p. I9
   Thin and brittle; rough sculpture.

5. *Atrina serrata* × ½, I view: **SAW–TOOTHED SEA PEN**
   p. I9
   Thin and brittle; finer sculpture.

a small plate, or platform, where the muscle is attached. The color is grayish brown, the interior dull white, not pearly. Originally described as *Mytilopsis leucopheata,* this bivalve lives in the brackish waters of coves and inlets, from Chesapeake Bay to the West Indies.

Family Pholadomyidæ

THESE ARE deep-water pelecypods, well represented as fossils, particularly in the Jurassic rocks, but now nearly extinct. The shells are equivalve, gaping behind, and are thin and white in color. The sculpture consists of strong radiating ribs. The hinge shows one obscure tooth in each valve.

Genus *Panacea* Dall 1905

1 Species

PANACEA ARATA Verrill & Smith   p. 193
This is a thin-shelled bivalve of triangular shape, a little under two inches in length. The posterior end is long and bluntly pointed and the anterior end is very short and abruptly truncated. The beaks are prominent. The surface bears deep, rather wide concave radiating furrows, separated by elevated, sharp-edged ribs. At the extreme posterior end the ribs are small and crowded. The color is white. This is a rare deep-water shell, taken south of Martha's Vineyard, Massachusetts, and it has no popular name.

Family Periplomatidæ

THE SHELLS are white, small, and usually fragile. They are slightly gaping. They are called "spoon shells" from a low, spoon-shaped tooth (chondrophore) in each valve. There is a small, triangular prominence lying next to this tooth that is generally lost when the animal is removed from its shell.

Genus *Periploma* Schumacher 1817

8 Species

PERIPLOMA FRAGILIS Totten   (Fragile Spoon Shell)
                                                        p. 61
The fragile spoon shell is a delicate little clam, thin and fragile as its name tells us. The right valve is a little more convex than, and it projects a little beyond the margin of, the left one. The shape is somewhat oval, the anterior gradually rounded and the

posterior end narrowed and slightly truncated. An elevated, angular ridge extends from the beaks to the posterior margin. The surface is minutely wrinkled, and the color is pearly white. This species is seldom more than one-half inch in length. It lives in the sand well beyond the lowest tide limits, and good specimens are not often found on the beach. It may be collected from Labrador to New Jersey.

PERIPLOMA LEANUM Conrad   (Spoon Shell)   p. 61
This bivalve is sometimes listed under its subgeneric name of *Cochlodesma leanum*. It is a larger mollusk, well over an inch in length. It is pearly white in color, sometimes with a thin, yellowish periostracum. The left valve is almost flat, and rounded at both ends, while the right valve is convex, and somewhat truncate at the posterior end. A faint ridge proceeds from the beaks to the same end. The spoon-shaped tooth in the hinge is nearly horizontal. This species occurs from the Gulf of St. Lawrence to North Carolina. It inhabits sandy bottoms in moderately shallow water.

PERIPLOMA ANGULIFERA Philippi   (Angled Spoon Shell)   p. 193
This species is about three-fourths of an inch long, with the anterior end longer than that of the last species, and less sloping. The color is white, and the surface is smooth. The chondrophore is relatively large, and points backwards. This bivalve occurs from Georgia to Texas.

Family Thraciidæ

THIS GROUP contains some rather large pelecypods, and some that are quite small. The valves are very inequal, and more or less gaping at each end. The beaks are prominent, and one of them is usually perforated.

Genus *Thracia* Blainville 1824

9 Species

THRACIA CONRADI Couthouy   (Conrad's Thracia)
p. 108
This is a large and peculiar bivalve found from Labrador to North Carolina. The shell is oval and inflated, the anterior end regularly rounded, and the posterior end narrowed and truncated. It is thin, light, and rather fragile, and dingy white in color, with the interior chalky white. Beneath its dull exterior

THE OYSTERS

1. *Pedalion alata* × 1, 2 views: **WINGED TREE OYSTER**
                                                          p. 20
     Pearly at center of interior.

2. *Anomia aculeata* × 1, 2 views: **PRICKLY JINGLE SHELL**
                                                          p. 34
     Thin-shelled; thorny.

3. *Ostrea frons* × ½, 3 views: **COON OYSTER**   p. 22
     Valves lobed.

4. *Anomia simplex* × 1, 3 views: **JINGLE SHELL**   p. 31
     Thin-shelled; waxy luster.

5. *Ostrea permollis* × 1, 2 views: **SOFT OYSTER**   p. 22
     Commonly fanlike.

6. *Ostrea virginica* × ½, 2 views: **VIRGINIA OYSTER**   p. 21
     Heavy and ponderous.

7. Young oysters, 30 days old, × 1

8. *Plicatula gibbosa* × 1, 2 views: **CAT'S–PAW**   p. 23
     Heavy folds; fan-shaped.

Plate 10   45

THE SCALLOPS

1. *Pecten irradians* × 1, 2 views: **BAY SCALLOP**   p. 23
   Variable, but usually not brightly colored.

2. *Pecten gibbus amplicostatus* × 1, 1 view   p. 24
   White ventral, mottled dorsal.

3. *Pecten raveneli* × 1, 2 views: **RAVENEL'S SCALLOP**   p. 25
   Strongly ribbed; commonly pinkish.

4. *Pecten gibbus* × 1, 1 view: **CALICO SCALLOP**   p. 24
   Gaily colored; very variable.

5. *Pecten ornatus* × 1, 1 view: **ORNATE SCALLOP**   p. 25
   Occasional unspotted ribs.

6. *Pecten grandis* × ⅔, 1 view: **DEEP–SEA SCALLOP**   p. 26
   Large, pinkish; nearly orbicular.

there is a pearly layer. The clam gets to be nearly four inches long. The beaks are about central and turned a little backward, and the peculiar feature is that the beak of one valve, the right one, is perforated to receive the point of the other beak. The surface is coarsely wrinkled by growth lines, undulated by a prominent ridge extending from the beaks to the posterior angle. The right valve is larger and more convex than the left, and it projects a little beyond it. The hinge is toothless, and there is a thin brownish periostracum. This mollusk lives in the sands just below the low-water mark, and is not especially abundant over most of its range, the majority of specimens being taken off the New England coast.

Family Pandoridæ

SHELLS SMALL, inequivalve, very thin and flat, with inconspicuous beaks. The color is white, and the interior is very pearly. These clams are distributed in all seas, and prefer rocky bottoms.

Genus *Pandora* Hwass 1795

7 Species

PANDORA GOULDIANA Dall   (Gould's Pandora)   p. 92
This little clam is one inch or less in length and grayish white in color. The shell is irregularly wedge-shaped, rounded before, with a recurved, truncated tip behind. The mollusk is exceedingly thin and flat. The anterior portion of the basal margin is strongly curved. Three distinct lines radiate from the beaks. The two valves are unsymmetrical, the right one flat and the left one slightly convex. The interior is pearly and highly polished. This curious little bivalve is promptly recognized by its upward-curved tip, its iridescent interior, and its remarkable thinness. It seems scarcely possible that there is room enough for an animal between its compressed valves. It frequents rocky and stony bottoms and is generally fairly plentiful on oyster beds. Single valves are often found in the drift of flotsam and jetsam that marks the limits of the tides. The range is from Prince Edward Island to North Carolina.

PANDORA TRILINEATA Say   (Three-lined Pandora)
p. 92
This name has often been applied to the shell just described, but it belongs to a much more slender and graceful species living from the Carolinas to Florida and Texas. The length is about one inch, and the color is white, with a very pearly inner surface. Its station in life is the same as its northern relative, rocky and pebbly bottoms in moderately shallow water, and it may be dis-

tinguished from *gouldiana* by the fact that the shell is more than twice as long as it is high, with a marked concave area between the beaks and the upturned posterior tip. A species that very closely resembles *gouldiana* but is noticeably larger, is sometimes found from Massachusetts to Newfoundland. This is *Pandora glacialis* Leach (p. 92).

Family Lyonsidæ

SHELL SMALL, inequivalve, and fragile. The hinge bears a narrow ledge to which the ligament is attached. The interior is pearly. These mollusks are represented in both the Atlantic and Pacific Oceans, living in shallow water as a rule.

Genus *Lyonsia* Turton 1822

7 Species

LYONSIA HYALINA Conrad   (Glassy Lyonsia)   p. 92
The glassy lyonsia is an attractive little bivalve, about three quarters of an inch long. The shell is thin and fragile, translucent and pearly. The anterior end is rounded, with the posterior end elongated, narrowed, and slightly truncate. The beaks are prominent. The surface is covered with radiating wrinkles that are minutely fringed so as to entangle grains of sand, with which the margins of the shell are often coated. An enlarged view shows this on plate 15. This species occurs all along the Atlantic coast, preferring shallow waters and sand bottoms. The delicate valves are commonly washed ashore, but they can stand but little rolling about in the surf, and perfect specimens are likely to be few and far between.

LYONSIA FLORIDANA Conrad   (Florida Lyonsia) p. 193
This is a trim little shell quite like *hyalina* but narrower and more pointed. The valves are fragile and translucent, and somewhat pearly, with a paper-thin periostracum. The anterior end is sharply rounded, with the posterior end prolonged and truncate. The shell is quite inequivalve, and the surface bears widely spaced radiating lines. This attractive little clam lives in shallow water on the west coast of Florida, and ranges south to Texas.

LYONSIA BEANA Orbigny   (Gaping Lyonsia)   p. 193
This is a larger species, about an inch long, that is very inequivalve, and rather unpredictable in shape. The posterior end is prolonged and gaping, and one valve is usually quite concave with the other more flat. The color is shiny white, often polished. This species lives in fairly deep water, and is found from North Carolina to the West Indian Islands.

Genus *Lyonsiella* Sars 1872

6 Species

LYONSIELLA GRANULIFERA Verrill   p. 193
This shell is higher than long, and somewhat triangular in shape.
The beaks are prominent, centrally situated, and turned forward.
The dorsal margin is convex behlnd the beaks, and concave in
front of them.   The surface is smooth and the color white, the
interior pearly.   Less than one inch long, this shell occurs in
deep water, off Chesapeake Bay.

Family Poromyidæ

SHELLS SUBEQUILATERAL, and inequivalve, thin, and somewhat
pearly within. Surface commonly granulated. These are small,
widely distributed bivalves, chiefly in deep water.

Genus *Poromya* Forbes 1844

9 Species and varieties

POROMYA SUBLEVIS Verrill   p. 193
This is a short and high shell, about three-fourths of an inch long.
The valves are well inflated, with the beaks large and curved
inward and forward. Both ends are short and bluntly rounded,
the posterior one rather oblique. The surface is smooth, with
very minute radiating lines. The color is white, with a thin
yellowish periostracum, and the interior is polished white. This
species lives in deep water off the Delaware coast, and does not
appear to have a common name.

Family Cuspidariidæ

THESE ARE small, pear-shaped bivalves, mostly confined to deep
water. They are called "dipper shells" from their elongated handle-
like posterior ends.

Genus *Cuspidaria* Nardo 1840

28 Species and varieties

CUSPIDARIA GLACIALIS Sars   (Northern Dipper)
                                        p. 193
This little bivalve is a fair-sized representative of the "dipper
shells," attaining a length of about three-fourths of an inch.
The posterior end is narrowed to form a definite "handle," but

it is not as long, or as pronounced, as with some of the group, so that the shell appears to be a little squat. The beaks are high, and the anterior end is regularly rounded. Grayish white in color, this species is one of the commonest of its group, and may be obtained in moderately deep water from Canada to Florida.

CUSPIDARIA MEDIA Verrill & Bush (Common Dipper)
p. 193

About one-half inch in length, this species resembles the last, *gracilis*, but it is smaller, and decidedly more swollen, the posterior end more narrow and proportionally longer. The anterior end is evenly rounded, and the beaks are rather prominent. The surface is smooth, bearing very minute lines of growth, and the color is yellowish white. This species is not uncommon south of Martha's Vineyard, in rather deep water.

CUSPIDARIA ROSTRATA Spengler (Rostrate Dipper)
p. 193

This dipper shell is almost one inch in length, and it boasts one of the most pronounced "handles" of any member of its group. The valves are well inflated, with the anterior end rather sharply rounded and the posterior end abruptly narrowed and drawn out to form a relatively long rostrum. The surface bears delicate lines of growth which form strong transverse wrinkles on this rostrum. Yellowish white in color, and shiny white within, this interesting little pelecypod ranges from the Arctic Ocean to the West Indies, occurring in deep water.

CUSPIDARIA MICRORHINA Dall (Long-handled Dipper)
per) p. 193

This species is very much like the last, but it is a larger bivalve, averaging about one and one-quarter inches long, and its valves are less inflated. The shape is the same, with the rostrum proportionally shorter but thicker. This species, which used to be considered a variety of *rostrata* but is now accorded full specific rank, lives in deep water off the coast of Florida.

CUSPIDARIA COSTELLATA Deshayes (Costellate Dipper)
per) p. 92

This is a little fellow, about one-fourth inch long. The beaks are nearly central, and the anterior end is regularly rounded, with the posterior end produced into a narrow rostrum. A series of strong radiating ribs marks the anterior half of the shell, which is scalloped at the margin. The color is bluish white, and the surface is shiny. This bivalve lives in the mud in deep water, off southern Florida.

CUSPIDARIA GLYPTA Bush (Sculptured Dipper) p. 193
This is an ornate little fellow, scarcely more than one-eighth of

an inch in length. The outline is about the same as the others of the group, with the anterior end rounded and the posterior rostrate. The surface is strongly sculptured with broad radiating folds. The color is pale brown. This is another deep water clam, ranging from Cape Hatteras to the West Indies.

Genus *Myonera* Dall & Smith 1886

7 Species

MYONERA GIGANTEA Verrill   p. 193
This species is shaped much like a "dipper shell," but the rostrum is short and pointed. The valves are thin and delicate, and considerably inflated. The surface bears crowded growth lines. The color is white, and there is a thin grayish periostracum. This is a rare shell, found in deep water (nearly 2000 fathoms) off the Virginia coast.

Family Verticordiidæ

THESE ARE MOSTLY small mollusks, very pearly inside. The shells are equivalve, or nearly so, well inflated, and quite solid in substance. They are chiefly dwellers in deep water.

Genus *Halicardia* Dall 1894

1 Species

HALICARDIA FLEXUOSA Verrill & Bush   p. 193
This is an obliquely swollen shell, about one and one-half inches in height, and nearly the same in length. The beaks are considerably incurved and turned forward. The outline of the shell is very flexuous, owing to slopes that divide the surface into three distinct areas. The anterior end is the shorter. The color is grayish white, with the interior more or less pearly. This is an uncommon species, living in deep water off the New England coast.

Family Pleurophoridæ

SHELLS LARGE AND THICK, almost circular in outline. There is no lunule. Natives of cold seas, the periostracum is thick and wrinkled.

Genus *Cyprina* Lamarck 1818

1 Species

CYPRINA ISLANDICA Linne   (Black Clam)   p. 96
The black clam is a large and robust pelecypod, four inches long

when fully grown. The shell is thick and heavy, and roughly circular in outline. The beaks are elevated, and turned forward and inward so as to come nearly in contact. The periostracum is black or deep brown, coarse, shiny, and rough with crowded and loose wrinkles. The interior is white. This is a northern species, found from the Arctic Ocean to Cape Hatteras. It is sometimes confused with the common quahog, *Venus mercenaria*, but it may be distinguished by its periostracum, and by the lack of a purplish border along the interior margin of the shell. The black clam appears to be most abundant offshore where large rivers empty into the ocean, and in such places their valves are sometimes washed ashore in considerable numbers. Young specimens, brownish tan in color, are common in the stomachs of fishes. This species is not very abundant south of the New England states.

Family Cyrenidæ

THESE ARE MOLLUSKS of brackish or semi-fresh waters. The shell is somewhat oval, and there is a rough periostracum, often eroded in places. They inhabit warm and temperate seas.

Genus *Polymesoda* Rafinesque 1820
4 Species and varieties

POLYMESODA CAROLINENSIS Bose   (Carolina Polymesoda)   p. 92

This bivalve is about one and one-half inches in length, and at first glance it looks much like a typical river clam. The shell is oval, considerably swollen, and covered with a shining green periostracum. The hinge is weak, and the interior is white. The valves are generally more or less eroded in the neighborhood of the beaks, owing to the mollusk's preference for brackish waters where acids are apt to be present. These greenish clams are generally quite abundant in tidal marshes and river-fed shallow lagoons, all the way from the Carolinas to Texas.

POLYMESODA FLORIDANA Conrad   (Florida Polymesoda)   p. 92

This is a smaller species, usually less than in inch in length. Its color is purplish white, darker at the margins. The shell is oval, rather thin but sturdy, with the beaks moderately prominent. The anterior end is rounded, and the posterior end is slightly prolonged. The hinge structure is weak, and there is a thin but rough periostracum. This little bivalve is an inhabitant of mangrove swamps, tidal marshes, and brackish waters generally. It ranges from Florida to Texas, and is usually abundant in favorable locations.

Family Astartidæ

SMALL, BROWNISH pelecypods, usually sculptured with concentric furrows. The ligament is external. The soft parts are commonly brightly colored. There are many species, distributed largely in cool seas. This is a confusing group. The species making up the genus *Astarte* are all much alike, but at the same time they show some variation, and many of them have been named and renamed until we find the same shell listed under several different names in older conchological literature. Take for example the species *Astarte borealis*. This pelecypod was named *borealis* by Schumacher in 1817. It was named *semisulcata* by Leach in 1819, *veneriformis* by Wood in 1828, *lactea* by Broderip in 1874, *richardsonii* by Reeve in 1855, *withami* by Smith in 1839, *producta* by Sowerby in 1874, *placenta* by Morch in 1883, and *rhomboidalis* by Leche in 1883! Its correct name is *Astarte borealis* Schumacher.

Genus *Astarte* Sowerby 1816

22 Species and varieties

ASTARTE CASTANEA Say   (Chestnut Astarte)   p. 92
This is a smooth astarte, with a shell that is small but thick and solid, with a strong hinge and broad hinge line. The outline is somewhat kidney-shaped, with the beaks nearly central and considerably elevated. About one inch long, the surface bears numerous concentric wrinkles, but it lacks the deeper furrows that are so characteristic of most of this group, or has them only weakly defined. The shell is covered with a rich chestnut-brown periostracum, often eroded near the beaks. The interior is shiny white. This is a mud-burrowing clam, living from Maine to New Jersey. In life the foot of the animal is bright vermilion, and when seen protruding from the partly open valves in shallow water presents an extremely colorful sight.

ASTARTE UNDATA Gould   (Waved Astarte)   p. 92
The waved astarte is a mahogany-brown little clam, one and one-quarter inches long. The shell is robust and roughly triangular. The posterior slope is rather straight, while the anterior slope bears a long, deeply excavated lunule. The beaks are elevated and pointed. The surface is decorated with about fifteen strongly developed concentric ridges and furrows, the furrows being widest and strongest at the center of the valve and vanishing at each end. There is a thick and glossy reddish-brown periostracum, and the interior of the shell is polished white. The waved astarte may be found from New England to Cape Hatteras, generally in small colonies, in mud a few feet

beyond the low water limits. Young shells are frequently found in marine fishes.

ASTARTE SUBÆQUILATERA Sowerby   (Lentil Astarte)

p. 92

Formerly called *Astarte lens* Stimpson, this shell is slightly more than one inch in length and moderately convex, with prominent beaks. The anterior slope is a bit concave, and the posterior end is broadly rounded. There are some fifteen squarish concentric ridges, more or less obsolete toward the posterior end. The margin is finely crenulate within. The periostracum is yellowish brown. This species lives in deeper water, and has a wide distribution, being reported from Labrador to Florida.

ASTARTE STRIATA Leach   (Striate Astarte)   p. 92

This is a small example, about one-half inch long. The shell is oval-triangular, and moderately stout. The beaks are prominent, pointing forward, and the lunule is broad and deeply excavated. The surface is marked by numerous closely spaced concentric ridges that are not elevated. The periostracum is dark brown. This small species lives from Massachusetts north to Greenland.

Family Gouldiidæ

SMALL, TRIANGULAR SHELLS, solid in substance. Lunule distinct. Strong hinge structure. Shallow to moderately deep water.

Genus *Gouldia* Adams 1847

2 Species

GOULDIA MACTRACEA Linsley   p. 193

This is a small but rugged shell, about one-third of an inch in length. Its shape is triangular, with the beaks forming the apex. The anterior and posterior slopes are pronounced, and the basal margin is evenly rounded. The hinge is thick and sturdy. The surface has a few undulating concentric waves, often rather indistinct, and very minute radiating striæ. The color is yellowish green. This little bivalve lives in relatively shallow water, from Massachusetts to Florida.

Family Carditidæ

SMALL, GENERALLY SOLID shells, equivalve, and usually strongly ribbed. There is an erect, robust tooth under the umbones. These pelecypods are found in warm, temperate, and cold seas.

Genus *Cardita* Brugière 1792

3 Species

CARDITA FLORIDANA Conrad (Bird Shell) p. 28
The bird shell is about an inch in length, sometimes a little longer, and it is colored yellowish white, blotched with purple and brown. Old individuals may be unspotted. The shell is heavy and solid, and bluntly oval in outline. There are about fifteen robust radiating ribs, with raised scales upon them. The interior is porcellaneous, and pure white. The colorful little bird shell is an inhabitant of bays, coves, and similar protected bodies of water, from Florida to Texas. Specimens can usually be found in the mud a few feet beyond the low-water line, while empty valves are generally plentiful in the drift that marks the high-tide limit. Many thousands of shells are used annually in the manufacture of shell novelties.

CARDITA GRACILIS Shuttleworth p. 92
This clam, which never seems to have acquired a popular name, is about one and one-half inches in length, and mottled gray and purplish brown in color. The shell is elongate, wedge-shaped, and moderately solid. The anterior end is short and rounded, and the posterior end is broad and sloping. There are about fifteen stout radiating ribs, crossed by fine growth lines. This graceful shell occurs on the Florida coast, but it is not overly abundant. Fresh valves are quite colorful, the inner surface deep purple.

Genus *Venericardia* Lamarck 1801

6 Species and varieties

VENERICARDIA BOREALIS Conrad (Northern Cardita)
p. 92
The northern cardita is about one inch in length and grayish white in color, with a brownish periostracum. The shell is very thick and solid, with the beaks elevated and incurved, rendering the shell heart-shaped when viewed from the end. There are about twenty radiating ribs, wider than the spaces between them. The interior is glossy white. This is a cold-water pelecypod, ranging from the Arctic Ocean to about the vicinity of New Jersey on the east coast, and to Oregon on the west coast. In life the ribs are often practically concealed by the thick and shaggy periostracum.

Family Chamidæ

SHELLS THICK and heavy, irregular, inequivalve. They are attached to some solid object, the fixed valve being the larger and more convex. They are natives of tropical and subtropical seas.

Genus *Chama* Linne 1758

4 Species

CHAMA MACEROPHYLLA Gmelin   (Jewel Box)   p. 12
The jewel box is a thick and ponderous oyster-like bivalve about three inches in length. Its color varies from pink and rose to yellow. The shell is irregularly rounded in outline, and is found attached to stones, corals, and other shells, or in the cracks and crevices of rocks or coral growths. The adhering valve is the larger and deeper of the two. The surface is sculptured with many distinct scalelike foliations, and the margins are finely crenulate. These bivalves inhabit warm or tropical seas, commonly living in crevices and other cramped quarters, and they are so firmly fixed that a hammer and chisel are a necessary part of the collector's equipment. The jewel box is a gregarious species, found in rather deep water off the Florida coast.

CHAMA SARDA Reeve   (Little Jewel Box)   p. 41
This is a smaller species, generally averaging about one inch in length. The shell while small is robust, with the surface bearing many wavy, tubular scales, arranged like shingles, some of them overhanging the margin, and often frondlike in character. The lower valve is white, the upper with reddish rays. This species also lives attached to some bottom object by its larger and deeper lower valve. Its home is in the Florida Keys.

Genus *Echinochama* Fischer 1887

1 Species

ECHINOCHAMA ARCINELLA Linne   (Spiny Chama)
p. 12
This is a small bivalve, about one and one-half inches long. The shell is robust and inflated, with the beaks curved forward. There are some seven or eight strong ribs on each valve, spread fanwise, and each with erect tubular spines throughout its length. The surface between the ribs is covered with beadlike pustules. The color is white, often reddish or purplish within. The spiny chama is an odd little shell, common from North Carolina to the West Indies. It is attached to some solid object in its youth, and later in life it becomes free, but the attachment

scar, in the form of a smooth area, is always present and visible just in front of the umbo on the right valve. Specimens may be obtained by dredging just off shore, the collector frequently finding several juveniles attached to each other, forming clusters. Single valves can usually be picked up on the beach, but unless they are fairly fresh the spines are likely to be broken and worn.

Family Lucinidæ

SHELLS GENERALLY ROUND, compressed, and equivalve, with beaks that are small but definite. Members of this family are distributed in tropical and subtropical seas for the most part, and are usually white in color.

Genus *Lucina* Bruguière 1797

14 Species

LUCINA PENNSYLVANICA Linne   (Pennsylvania Lucine)

p. 93

This lucine is about two inches in length. The shell is nearly circular in outline, and moderately well inflated, with beaks that are inclined forward. There is a deep fold running from the beak to the posterior margin. The surface is marked with widely separated, sharp, concentric ridges. The color is white, with a pale brownish, thin periostracum. This distinctive pelecypod occurs from Cape Hatteras south, living offshore in moderately deep water. The peculiar posterior fold, a characteristic of this group, gives the appearance of one shell cupped within another. Single valves are not uncommon on southern shores, and when collected fresh, with the periostracum intact, they are very pretty shells.

LUCINA JAMAICENSIS Lamarck   (Jamaica Lucine)

p. 93

This species is also about two inches in length. The shell is heavy and solid, and roughly circular in outline. The characteristic fold extends along the posterior margin. The surface bears many fine concentric ridges, rather widely spaced, and the color is pale yellowish white. This species is not as inflated as the last, nor are its concentric ridges as sharp. It is a shallow water clam, thriving on sand bars close to shore. It is fairly common in Florida, and ranges well down into South America.

LUCINA LEUCOCYMA Dall   (Sulcate Lucine)   p. 193

This is a small but sturdy shell, a little less than one-half inch in length, and a bit more than that in height. The shape is somewhat triangular, with prominent beaks. The shell bears five broad folds, or radiating scorings, producing a lobed effect to the outline. The color is pure white, and the only sculpture consists of fine concentric lines. The inner margins are finely crenulate. This species lives in fairly deep water, and occurs from North Carolina to the West Indies.

LUCINA FLORIDANA Conrad   (Florida Lucine)   p. 93
This is a yellowish-white shell, about one and one-half inches long. The valves are thick and solid, circular in outline, and considerably compressed. The beaks are small but prominent, and inclined to turn forward. The fold from beak to posterior margin is less conspicuous than with most of this genus. The surface bears fine concentric growth lines, and there is a thin, yellowish periostracum. This clam is an inhabitant of the Gulf of Mexico, living from Texas to the west coast of Florida. Its station is just offshore in shallow water, and shells can generally be found along the beaches throughout its range. It may be distinguished from the other members of its genus by its relative thinness and by its rather indistinct posterior fold.

LUCINA FILOSA Stimpson   (Northern Lucine)   p. 193
This is an attractive shell, nearly two inches in length. The shape is rather circular, with the hinge line nearly straight. The valves are compressed. The beaks are small but prominent, and turned forward. The color is white, and the shell is decorated with sharp concentric ridges that are rather widely separated. This is not a very common shell, but it may be found all the way from Maine to Florida. Unlike most of this group, it is a cold-water species, and it is found at increasingly greater depths as it passes southward.

Genus *Divaricella* Martens

2 Species

DIVARICELLA QUADRISULCATA Orbigny   pp. 12, 193
This is an odd little shell, ivory-white in color and about an inch in length. The shell is moderately solid, circular in outline, and rather plump. The surface is sculptured quite unlike any similar shell, with prominent grooves that are bent obliquely downward at both ends. The inner margins of the valves are minutely crenulate. This species enjoys a wide distribution, being found from New England to Brazil. It lives in rather deep water, but single valves are not uncommon upon beaches all along our coasts, especially in the south. Unless the shell is fairly fresh, however, the odd sculpture is apt to be partially or completely obliterated. *Divaricella dentata* Wood is a larger species, found from North Carolina to the West Indies. This form is also white, and carries the same sculpture as its smaller and more abundant relative.

Genus *Codakia* Scopoli 1777

7 Species and varieties

CODAKIA ORBICULARIS Linne   (Great White Lucine)

p. 93

The great white lucine is a handsome and showy shell, much prized by collectors. It used to be listed as *Lucina tigrina*. The length is about three inches, and the color is white, sometimes with a border of pinkish or lavender on the inside. The shell is large and solid, quite orbicular in outline, and but little inflated. The beaks are sharp and prominent, and the hinge teeth large and sturdy. The surface is marked with many narrow radiating ribs, crossed by elevated growth lines, giving the shell a cross-ribbed appearance. This bivalve may be found sparingly from North Carolina to the West Indies. It is quite abundant in the Florida Keys, where it lives in the sand in shallow water.

CODAKIA ORBICULATA Montagu   (Little White Lucine)

p. 93

The little white lucine is much like a small edition of the species just described. It is only about one inch in length, and pure white in color. The shell is sturdy, and circular in outline, with the beaks placed near the anterior end. The surface has broad and distinct radiating ribs which are crossed by numerous fine concentric lines, as well as frequent ridges marking rest periods in shell growth. The little white lucine ranges from the Carolinas to the West Indies, and is partial to shallow water, preferably where there is a sandy bottom.

Genus *Loripinus* Monterosato 1883

2 Species

LORIPINUS CHRYSOSTOMA Philippi (Buttercup)   p. 28

This species is known by several popular names, chief among them being "Buttercup Shell" and "Apricot Shell." It is about two inches long, and chalky white in color, with the interior orange or bright yellow. The shell is strong, considerably inflated, with rounding margins, and the beaks are low but prominent. The surface appears to be smooth, but there are numerous very faint growth lines. The ligament is bright red in living specimens. The colorful buttercup shell is well known to visitors in the southland, where it is gathered in quantities for making decorative shell novelties. It is at home in the tidal flats of coves, bays, and other protected bodies of water, from North Carolina to Cuba. The shells are usually very common on the beach, but, since the hinge is very weak, it is unusual to find both valves together.

LORIPINUS SCHRAMMI Crosse (Schramm's Loripinus) p. 93
This is a larger species, sometimes confused with the last. It is three inches or so in length and dull chalky white in color. The shell is strong and solid and quite globose, with well rounded, centrally placed beaks. The surface is decorated with small but sharp concentric lines. It lacks the brightly hued interior that makes the buttercup shell so attractive. This species used to go under the name of *Loripinus philippiana*. It may be found, sparingly, from Cape Hatteras to Mexico.

Genus *Myrtæa* Turton 1822

3 Species

MYRTÆA LENS Verrill & Smith p. 200
Like many of our small varieties, this one does not appear to have any popular name. It is a roundish shell with low beaks and very compressed valves. The color is dull white, with a greenish-gray periostracum, and the length is about a half inch. This species was originally described as *Loripes lens*. It is not particularly rare, and occurs from Cape Cod to South America, living in moderately deep water.

Family Cardiidæ

SHELLS EQUIVALVE and heart-shaped, frequently gaping at one end. The margins of the shell are serrate or scalloped. Native to all seas. This family contains the cockles, or heart-clams. In Europe these bivalves are regularly eaten, and "cockle-gathering" is a recognized seaside industry, but they are not used as food to any extent in this country. For many years nearly all of the shells belonging to this family were referred to the genus *Cardium*, and those found on our Atlantic coast will be found listed as *Cardium isocardia*, *Cardium islandicum*, etc. It is now recognized, however, that the genus *Cardium* should be restricted to those forms living in the Eastern Atlantic, and so our species have had their subgeneric names elevated to full generic rank.

Genus *Trachycardium* Morch 1853

2 Species

TRACHYCARDIUM EGMONTIANUM Shuttleworth
(China Cockle) p. 93
This bivalve has long gone under the name of *isocardia*, but it is now believed that the shell found along our southern shores is distinct from that species. The two forms are very closely related, but *isocardia*, now restricted to the Caribbean area, is

THE SCALLOPS AND THE FILE SHELLS

1. *Pecten dalli* × 1, 2 views: **DALL'S SCALLOP**   p. 27
Ribs on interior.

2. *Pecten ziczac* (upper valve) × 1, 1 view: **SHARP–TURN SCALLOP**   p. 25
Zigzag lines on flat valve.

3. *Abra æqualis* × 1, 2 views   p. 81
See text.

4. *Lima tenera* × 1, 2 views: **DELICATE FILE SHELL**   p. 30
Satinlike finish.

5. *Lima lima* × 1, 2 views: **FILE SHELL**   p. 30
Ribs with sharp scales.

6. *Lima inflata* × 1, 2 views: **INFLATED FILE SHELL**   p. 30
More inflated than others.

7. *Pecten islandicus* × ⅔, 1 view: **ICELAND SCALLOP**   p. 26
Large, with cordlike ribs.

Plate 12   61

THE MUSSELS

1. *Periploma fragilis* × 1, 2 views: **FRAGILE SPOON SHELL**
   p. 42
   Spoonlike structure under beak.

2. *Periploma leanum* × 1, 2 views: **SPOON SHELL**   p. 43
   Like above, but larger

3. *Mytilus exustus* × 1, 2 views: **SCORCHED MUSSEL** p. 35
   Brown; "lopsided."

4. *Crenella decussata* × 1, 2 views: **LITTLE ROUND MUSSEL**
   p. 39
   Nearly globular; finely striate.

5. *Modiolaria nigra* × 1, 1 view: **LITTLE BLACK MUSSEL**
   p. 38
   Dark; radiating lines at each end.

6. *Volsella tulipus* × ½, 2 views: **TULIP MUSSEL**   p. 36
   Thin-shelled; inflated.

7. *Lithophaga antillarum* × 1, 2 views: **ANTILLARIAN DATE**
   p. 38
   Elongate; cylindrical.

8. *Lithophaga nigra* × 1, 2 views: **BLACK DATE**   p. 38
   Like 7, but smaller and darker.

9. *Volsella modiolus* × ½, 1 view: **HORSE MUSSEL**   p. 35
   Heavy and solid.

10. *Volsella plicatulus* × 1, 1 view: **RIBBED MUSSEL**   p. 36
    Strong radiating ribs.

11. *Mytilus edulis* × 1, 1 view: **BLUE MUSSEL**   p. 34
    Smooth, blue to blue black.

a slightly larger shell, with more ribs, and more pronounced scales, so that it is a more spinose shell. The bivalve found from New Jersey to Florida is *T. egmontianum.*

Known popularly as the china cockle, the shell is from two to two and one-half inches long, and yellowish to creamy white in color. The interior is reddish purple. The valves are well inflated, rather thin, and oval, and the moderately prominent beaks are nearly central in position. The surface is sculptured with deeply chiseled radiating ribs, the tips of which are studded with sharp, recurving scales, most pronounced on the anterior and posterior slopes. The china cockle lives in the sand well below the low-water mark, often forming large-sized beds in sheltered situations. The shell is very popular with seashore visitors, its graceful lines and delicate coloring making it suitable for all sorts of fancy articles, such as pin cushions, ash trays, shell flowers, etc.

TRACHYCARDIUM MURICATUM Linne   (Common Cockle)   p. 12

The so-called common cockle attains a length of about three inches and is yellowish white in color, sometimes lightly speckled with brown, especially on the umbones. The interior is yellow. The shell is rounded and inflated, with valves that are equal in size, and heart-shaped when viewed endwise. There are from thirty to forty pronounced ribs, with about a dozen of the central ones almost or quite smooth over the umbonal region, the others crossed by erect, sharp scales. The margins of the valves are serrate and interlocking. The common cockle is found from North Carolina to South America. It prefers a moderate depth of water and a sandy bottom, but shells are washed up on the beach with nearly every incoming tide. This species may be recognized from the last, *egmontianum*, by its rounded, less oval shape, and by the color of its interior, which is yellowish instead of reddish.

Genus *Dinocardium* Dall 1900

1 Species and 1 variety

DINOCARDIUM ROBUSTUM Solander   (Great Heart Cockle)   p. 12

The great heart cockle is our largest member of this group, averaging between three and six inches in length. Its color is yellowish brown, irregularly spotted with chestnut and purplish marks. The posterior slope is brownish purple, and the shell's interior is salmon-pink. The shell is large and considerably inflated, with the posterior area flattened, dark, and polished. The beaks are strongly rounded. There are about thirty-five robust, flat ribs, regularly arranged. The margins of the valves are serrate. The great heart cockle is easily recognized by its dark and flattened posterior end, as well as by its size and shape. It prefers a sandy bottom and a moderate depth of water, and frequently forms large beds in estuaries. It occurs from Virginia to Brazil. A form living on the west coast of Florida is generally larger in size, more triangular in outline, and brighter in color. This form has been named *Dinocardium robustum* var. *vanhyningi* Clench & Smith.

Genus *Clinocardium* Keen 1936

1 Species

CLINOCARDIUM CILIATUM Fabricius   (Iceland Cockle)
p. 96

This is the Iceland cockle, known for many years as *Cardium islandicum* Bruguière. It is about two and one-half inches long and dull white in color. The shell is large and well inflated, with prominent beaks. The anterior end is a little shorter and narrower than the posterior. The surface bears about thirty-eight sharp-edged, radiating ribs, the furrows between them rounded and slightly wrinkled by lines of growth. The periostracum is stiff and fringelike, especially on young shells; in fact, one of the common names for this species is "hairy clam." The margins are scalloped, and the interior of the shell is straw-colored, sometimes bright orange in juvenile specimens. The Iceland cockle, as its name implies, is an Arctic species, ranging down our coast as far as Cape Cod. This bivalve is often found in the stomachs of cod and other marine fishes, and single valves are occasionally found on New England beaches.

Genus *Cerastoderma* Morch 1853

2 Species

CERASTODERMA PINNULATUM Conrad   (Little Cockle)
p. 96

This little cockle is only about one-half inch long. The shell is small, rather fragile, and nearly orbicular in outline, with a blunt ridge passing from the beaks to the posterior point of the shell. There are about twenty-five rounded, radiating ribs, on each of which is a series of scales, most pronounced near the margin. The color is dingy white, the interior flesh-colored in fresh specimens. This species is usually found with the young of the Iceland cockle, but they may be distinguished by the fewer ribs and the scales crossing them. The little cockle is an active pelecypod, said to scamper over the gravelly bottom with surprising speed, making expert use of its recurved, extensible foot. This species may be collected from Labrador to North Carolina.

CERASTODERMA ELEGANTULUM Beck   (Elegant Cockle)

p. 193

This little cockle is slightly larger than the last, and a little more coarsely ribbed. The color is white or yellowish white. There are from twenty-six to twenty-eight ribs, strongly roughened by closely spaced, arched and overlapping scales. This is a very rare shell in this country. It is found in northern Europe and in Greenland, and off shore in Labrador.

Genus *Papyridea* Swainson 1840

2 Species

PAPYRIDEA HIATUS Meuschen   (Spiny Paper Cockle)

p. 28

The spiny paper cockle is a common shell from the Carolinas to Brazil. It is about one inch in length, and white or pink in color, heavily mottled with rosy brown, inside as well as out. Occasional specimens are pure yellow. The shell is thin, compressed, somewhat elongated, and gaping at the posterior end. The beaks are low, and there are many fine, radiating ribs, smooth in the center of the shell but provided with short spines toward the extremities, sometimes overhanging the margins. This shell will be found in many older books under the name of *Papyridea spinosum*. It lives under aquatic plants in sandy situations, and single valves are common along our southern shores. It is a rather delicate shell, however, and seldom do we find both valves in place.

Genus *Trigoniocardia* Dall 1900

3 Species

TRIGONIOCARDIA MEDIUM Linne   (Oblique Cockle)

p. 12

This is a pretty little bivalve about one inch in length, and creamy white in color, more or less checkered with buff and purple. The shell is small but solid, and somewhat triangular in appearance, with the anterior margin regularly rounded and the posterior margin partially truncate, forming a distinct slope on that end. There is a sculpture of strong, rounded, radiating ribs. This is a neat fellow, quite brightly colored as a rule when taken freshly, but faded and dull in the majority of beach specimens. It lives just offshore in sandy situations from Cape Hatteras to the West Indies.

Genus *Lævicardium* Swainson 1840

2 Species

LÆVICARDIUM LÆVIGATUM Linne   (Egg-shell Cockle)

p. 12

This species is known as the "egg-shell cockle." It used to go under the name of *Lævicardium serratum*. It is about two inches high and ivory-white in color, more or less marked with rusty orange. Occasional specimens are found that when quite fresh are very brightly hued. The shell is thin and inflated, with the anterior end curved a little more than the posterior. The beaks are small, and the surface is quite smooth, but with very faint traces of ribs. The inner margin is crenulate, and there is a thin, brownish periostracum. This bivalve may be found from North Carolina to the West Indies, living in moderately shallow water, preferably on a gravelly or pebbly bottom. The leaping ability of this clam is well established, and one collector reports a captive specimen making a successful getaway by using its powerful foot to leap from the boat.

LÆVICARDIUM MORTONI Conrad   (Morton's Cockle)

p. 93

Morton's cockle is a smaller form, seldom exceeding one inch in height. Its color is yellowish white, generally a little streaked with orange, and the interior is commonly bright yellow. As a rule the colors are more pronounced and lasting on this species than on the one just described. The shell is small, thin, and inflated, and obliquely oval in outline. The surface is smooth and polished, and the inner margins of the valves are crenulate. Morton's cockle extends nearly the length of the Atlantic coast-line, being known from Nova Scotia to Brazil. It is most abundant in the south, however, and southern specimens are generally the brightest in color, but even these fade somewhat as the shell dries out. This little pelecypod lives close to shore in water only a few feet deep.

Genus *Serripes* Beck 1841

1 Species

SERRIPES GRŒNLANDICUS Bruguière   (Greenland Cockle)   p. 108

This is a rather large clam, averaging some three or four inches in length. Its color is drab gray, with juvenile specimens sometimes showing a few zigzag darker lines. There is a thin grayish brown periostracum. The shell while large is not very thick, and it is but little inflated. The outline is somewhat triangular, with the beaks centrally located and slightly incurved. The anterior end is regularly rounded, while the posterior end is partially truncate and widely gaping. The surface bears numerous concentric lines, and several radiating ridges, most pronounced on the two ends. At first glance this shell resembles a *Spisula* or *Mactra*, to be described later, but one look at the hinge shows that it lacks the characteristic spoonlike cavity of these genera. It lives in fairly deep water, from Greenland to Cape Cod, but it is not very common on our shores.

Family Veneridæ

THIS IS THE LARGEST pelecypod family, and it has the greatest distribution, both in depth and in range. Named for the goddess Venus, the shells of this group are noted for their graceful lines and beauty of color and sculpture. The shells are equivalve, commonly oblong-oval in outline, and porcellaneous in texture. The mollusks are burrowers just beneath the surface of sand or mud, and are never fixed in one place. They are native to all seas, and since ancient times many of them have been used by man for both food and ornament.

Genus *Dosinia* Scopoli 1777

2 Species

DOSINIA DISCUS Reeve (Disk Shell) p. 96
This is a trim and neat shell, about two and one-half to three inches in length. It is large and moderately thin, circular in outline, and considerably compressed, with small but prominent beaks. The surface is decorated with numerous distinct concentric lines. The hinge is thick and sturdy. The color is glossy white, with a thin, yellowish periostracum. This distinctive pelecypod occurs from New Jersey to Texas, but it is considered an uncommon species in the northern parts of its range. The thin periostracum peels away easily, revealing a shiny white shell.

DOSINIA ELEGANS Conrad (Elegant Disk) p. 12
This is a very similar species, also pure white in color and up to three inches in length. Its surface bears numerous uniformly spaced concentric lines, but they are not so crowded as they are with *discus*. Living from Cape Hatteras to the West Indies, this bivalve is often confused with the last, but it is less compressed, and easily distinguished by the spacing of its concentric lines. Clench states that *elegans* has from eight to ten ribs to a centimeter, while *discus* has twenty to a centimeter. *Dosinia elegans* lives just offshore in moderately shallow water, and paired valves are often common on southern beaches.

Genus *Macrocallista* Meek 1876

2 Species

MACROCALLISTA NIMBOSA Solander   (Sun-ray Shell)

p. 28

This bivalve is large and showy, sometimes reaching a length of six inches. Its color is pinkish gray, with radiating lilac bands, and in fresh specimens the interior is salmon-pink. The shell is smooth, thick, porcellaneous, and elongate-oval in form, with depressed beaks. The anterior end is short and rounded, and the posterior end is elongate and rounded. The surface is glossy, but with very faint concentric and radiating striations. The inside is polished, with smooth margins. This is a handsome shell, but half-grown, two- or three-inch individuals are generally more brightly colored than old adults. The species is at home in sand bars just beyond the low-water mark, especially in bays and coves, from North Carolina to Cuba, and west to Alabama.

MACROCALLISTA MACULATA Linne   (Checkerboard)

p. 28

This species is two or three inches long, and roundish oval in outline, not elongate like the last one. The color is yellowish buff, with squarish spots of violet brown distributed over the whole shell, but with a couple of radiating rays usually present. The valves are thick and solid, and the surface is porcellaneous, with very fine radial lines. The interior is white and polished, the pattern sometimes showing through. The margins are smooth. The well-named checkerboard is another handsome species, and very popular with collectors. It inhabits shallow water where there is a sandy bottom, and is often turned out and eaten by gulls while the tide is out. Single valves are fairly common on the beaches in suitable locations. The species ranges from North Carolina to Brazil, but it is not very abundant on the east coast. It is much more plentiful on the Florida west coast.

Genus *Pitar* Romer 1857

8 Species

PITAR FULMINATA Menke   (Lightning Venus)   p. 13

This is a plump little bivalve, about an inch and a half in length. The shell is rounded oval, with small beaks, and with the posterior end a little longer than the anterior. There is a sculpture of very fine concentric lines. The interior is polished, and the margins are smooth. The color is white, with brown or orange spots commonly arranged in a zigzag pattern, and some ex-

amples are quite colorful. This pretty little shell occurs on
sandy beaches from the shore line to a depth of several fathoms,
all the way from Cape Hatteras to Brazil.

PITAR MORRHUANA Gould   p. 96
This is a dull white, sometimes rusty-colored clam, about two
inches in length. The shell is roundish oval, rather thin, with
the valves quite convex. The anterior end is about half the
length of the posterior. The margins are regularly rounded
behind and at the base. There is a heart-shaped lunule in front
of the moderately elevated beaks. The surface is smooth, with
small lines of growth, and the interior is white, sometimes
polished. This unattractive little bivalve occurs from Prince
Edward Island to North Carolina, living in the sands well below
the low-water level. The shell appears somewhat like a small
and wave-worn *Venus mercenaria*, but it is not as solid, has a
smaller tooth structure, and it lacks the purple border of the
latter clam. This species has had many names, and it may be
found listed as *Cytheria convexa, Callista convexa, Callista sayana*,
and *Callocardia convexa*. The true *convexa* is a very similar shell
that occurs only as a fossil.

PITAR DIONE Linne   (Elegant Venus)   p. 28
This is considered by many to be the handsomest bivalve shell
in the western Atlantic, but it very nearly misses our shores,
being found from Texas and the West Indies south. About two
inches long, the shell is plump, with rather prominent beaks.
The anterior end is broadly rounded, and the posterior end is
gently sloping. The surface bears many deeply cut concentric
grooves, and there is a more or less distinct ridge running in an
easy curve from the beaks to the posterior margin; this ridge
is provided with one or two rows of long spines. The color of
the whole shell, spines and all, is a delicate pinkish violet.
Formerly known as *Venus dione*, this is the shell used by Linnæus
in his original description of the genus *Venus* in 1758. It is a
strikingly beautiful shell, not as fragile as it appears, and it is
generally considered a prize in any collection.

Genus *Antigona* Schumacher 1817

5 Species

ANTIGONA LISTERI Gray   (Lister's Venus)   p. 96
This is a grayish white clam, about three inches in length. The
shell is thick and solid, and broadly oval in outline. It has a
sculpture of impressed lines, crossed by concentric ridges that
are both heavy and sharp. On the posterior end these ridges
often expand to form flaring, bladelike structures. This is a
rather uncommon West Indian bivalve, found occasionally as

far north as central Florida. It burrows in the sand well below
the low-water line, and in many ways it resembles the common
quahog, or round clam, but it can easily be identified by its
sharp, reticulated sculpture.

Genus *Chione* von Muhlfeld 1811

10 Species

CHIONE CANCELLATA Linne   (Cross-barred Venus)

p. 28

The cross-barred venus is about one and one-quarter inches long,
and cloudy yellowish white in color, decorated with zigzag or
triangular patches of purplish brown, sometimes in the form of
radiating bands. The interior is usually purple. The shell is
small, thick, and solid, the beaks elevated and situated well
forward. The surface is sculptured with a series of well-elevated
rounded ribs, which are crossed by concentric ridges of the same
size, forming a network of raised lines on the shell. The inner
margins are crenulate. This is a very attractive little shell, common on our southern beaches. Its station in life is the sand or
mud flats just beyond the low water mark. The cross-barred
venus is a typical Florida shell, ranging north as far as Cape
Hatteras, with the largest specimens coming from the more
northern parts of its range.

CHIONE INTERPURPUREA Conrad   (Mottled Venus)

p. 28

The little mottled venus is about one and one-half inches in
length, sometimes a little larger. Its color is yellowish buff,
heavily mottled with dark brown. The shell is strong and solid,
the ventral margins strongly convex. The beaks are prominent
but low. There are many crowded concentric ridges, strongly
wrinkled over the posterior half. The inner surface is smooth,
the margins crenulate. This sturdy little clam is somewhat
larger than the last one, and it lacks the strong concentric ridges
of that species. It occurs in fairly deep water and appears to
prefer gravelly bottoms. Its range is from Cape Hatteras to
Florida and along the Gulf coast to Honduras, but in most
localities it is considered uncommon.

CHIONE LATILIRATA Conrad   (Broad-ribbed Venus)

p. 28

The broad-ribbed venus gets to be nearly two inches in length,
but its average size is somewhat less. It is grayish white in
color, irregularly marked with lilac and brown. The shell is
solid and sturdy, well inflated, and roughly triangular in outline.
The surface bears a series of large and broadly rounded concentric ribs, with deep furrows between them. These ribs tend

to pinch out at the posterior end. The shell has a very high polish. This is another handsome bivalve, found (not very commonly) in southern Florida and the West Indies. It lives in moderately deep water. Its broad ribs and furrows, plus its high gloss, render this pelecypod easy to identify, but wave-worn specimens found on the beach are likely to have lost their polish. This species will be found in some books listed as *Chione paphia*.

Genus *Venus* Linne 1758
8 Species and varieties

VENUS MERCENARIA Linne   (Quahog)   p. 108
This bivalve has a good many popular names, among them round clam, cherrystone clam, hard-shelled clam, little neck clam, and quahog. It is five or six inches in length when fully mature, and dull grayish white in color. The shell is large, thick, and solid, and rather well inflated, with the beaks elevated and placed forward. The surface bears numerous closely spaced concentric lines, most conspicuous near the ends, the central portion of the valves being smoother. Around the umbonal region the concentric lines are rather widely spaced, this feature being especially noticeable in young shells. The interior is white with a dark violet border near the margin. This is the chief commercial clam of the east coast, ranking second only to the oyster in shellfish value. When half-grown it is the delicious "cherrystone" of banquet fame, said to have a flavor surpassing almost any other bivalve. When older it is less tender, but used extensively for chowders, bakes, etc. It was a favorite food of the coast Indians, as is attested by the many shell heaps of ancient vintage found scattered from Maine to Florida, and from the Red Men we get the name of "Quahog," by which this clam is known in New England. The noted "purple wampum" was made from the colored edge of the valves. It is not uncommon to find pearls under the mantle of this species, sometimes of fair size, but of little value. The range of the quahog is from Maine to Florida.

VENUS MERCENARIA NOTATA Say   p. 108
This is a subspecies of the last named. It averages three or four inches in length, and differs from the typical form by commonly showing a pattern of purplish brown scrawls on the outer surface, and in generally lacking the purplish border within. It occurs from New England to Florida, but it is rather uncommon in the north.

   A closely related form occurs in southern waters. This is *Venus campechiensis* Gmelin, a bivalve with a larger and thicker shell than the common quahog. This form may be found from

Chesapeake Bay south, and it is the common form along the Texas coast of the Gulf of Mexico.

Genus *Anomalocardia* Schumacher 1817

2 Species

ANOMALOCARDIA CUNEIMERIS Conrad (Pointed Venus) p. 108
The pointed venus is about one-half inch long, and varies from grayish white to green and brown in color, sometimes with dark markings. The shell is small, thin, and wedge-shaped — decidedly pointed at the posterior end. The beaks are slightly elevated, and the surface of the shell is decorated with a series of rounded concentric ribs. The surface is shiny, and the inside of the valves is colored pale lavender, with the margins crenulate. This little shell is common on sandy beaches from Florida to Mexico. It lives in sand bars just offshore, from whence the empty valves are washed up with each incoming tide. Varying considerably in color, the neatly triangular pointed venus lends itself particularly well to the making of shell knicknacks.

ANOMALOCARDIA BRASILIANA Gmelin (Little Striped Venus) pp. 13, 109
This is a pretty and graceful little shell, about three-fourths of an inch in length. It is pale gray in color, with violet bands and checks. The shell is triangular, drawn out to a point at the posterior end, and the surface is smooth, with small concentric ribs. The posterior angle is beaded in the upper portion. The inner margins are crenulate. This species extends a little further north than the last, the striped venus occurring from Cape Hatteras to Brazil. It is also quite variable in color pattern, and in form as well, the degree of "pinching out" at the back differing considerably in different individuals. Like its slightly smaller relative, it frequents sand bars a few feet below the low-water limits.

Genus *Liocyma* Dall 1870

1 Species

LIOCYMA FLUCTUOSA Gould p. 200
This is a small shell, about one-half inch long and oblong-oval in shape. The valves are thin, and but little inflated. The color is white, beneath a thin yellowish periostracum. The beaks are nearly central, with the anterior end shortest and broadest. Both ends are widely rounded. The surface bears about twenty concentric ridges which fade near the margins. This little clam occurs from Greenland to northern Maine, and does not seem to have any popular name.

Genus *Gemma* Deshayes 1853

3 Species and varieties

GEMMA GEMMA Totten   (Gem Shell)   p. 108
The gem shell, sometimes called the amethyst gem shell, is a
diminutive fellow, about one-fifth of an inch long. The tiny
shell is broadly triangular in shape, with the beaks about central.
The surface is shining, with minute, crowded, concentric lines.
The inner margin is crenulate. The color of the shell is pale
lavender, the inside white, shading to purplish near the posterior
end. This gemlike shell is found in great abundance on the
sandy shores of protected bays and coves. The early settlers in
Massachusetts sent boxes of them back to England as curiosities.
There was a time when it was believed that they were the young
of the quahog, or round clam. The species ranges from Labrador
to Cape Hatteras.

Genus *Parastarte* Conrad 1862

1 Species

PARASTARTE TRIQUETRA Conrad   p. 200
This is a rather solid little shell, generally less than one-fourth
inch long. Its shape is triangular, with the beaks elevated,
prominent, and situated at the apex of the triangle. The surface
is smooth, with minute concentric lines, and the color is white
with a tinge of purple. The margins are crenulate. This little
bivalve occurs in Florida, and like many of the small deep water
forms, it has never acquired a popular name.

Family Petricolidæ

SHELLS ELONGATE, gaping behind, and with a weak hinge. They are
burrowing mollusks, excavating cavities in clay, coral, limestone,
etc. The cavity is gradually enlarged until the clam attains adult
size. Members of this family are distributed in warm and temperate seas around the world.

Genus *Petricola* Lamarck 1801

3 Species and varieties

PETRICOLA PHOLADIFORMIS Lamarck   (False Angel
Wing)   p. 124
This shell is about two inches long when fully grown, and it is
chalky white in color. The shell is thin, much elongated, and
somewhat cylindrical. The anterior end is very short and

acutely rounded, while the posterior end is narrowed, elongate, and a little gaping. The beaks are elevated, and just in front of them is a well-defined lunule. The surface is marked with growth lines and also with strong radiating ribs. At the posterior end the ribs are crowded and faint, but at the anterior end they are large and widely spaced. This is a burrowing clam, bearing a striking resemblance to the large and showy "Angel's Wing." It is a very common species on muddy shores and in salt marshes, where it burrows horizontally into the banks between the low and high tide marks. It sometimes bores into stiff clay, or even into some of the softer rocks, such as limestone. Its range is from Prince Edward Island to the West Indies.

Genus *Rupellaria* Fleurian 1802

2 Species

RUPELLARIA TYPICUM Jonas p. 200

This is a plump and rough shell, rather elongate, about an inch and a half in length. The anterior end is well rounded, while the posterior end is longer and gaping. The surface bears radiating lines, fine and close near the two ends of the shell, with undulating lines of growth between them. The color is grayish white. This, too, is a burrowing clam, with habits quite like the last species. It lives commonly in the clay banks and softer rocks between tides, from North Carolina to the West Indies.

Genus *Coralliophaga* Blainville 1864

1 Species

CORALLIOPHAGA CORALLIOPHAGA Gmelin p. 200

This is a rather thin-shelled bivalve about one and one-half inches long. The shell is cylindrical, rounded at each end, and gaping a little posteriorly. The valves bear faint radiating lines, but the surface appears fairly smooth. The color is pale tan. This is another burrowing clam, as its shape suggests. It is often found in the burrows of some other mollusks, sometimes in company with the rightful owner. It occurs on the west coast of Florida, and ranges south to Texas.

Family Tellinidæ

THE SHELLS of this family are generally equivalve and rather compressed, the anterior end rounded and the posterior end more or less pointed. The animals are noted for the length of their siphons. Natives of all seas, several hundred species have been described, including many fossil forms, and within this group we find some of the most colorful, highly polished, and graceful of the bivalves.

Genus *Tellina* Linne 1799

35 Species and varieties

TELLINA LINEATA Turton (Rose Petal) p. 28

The rose petal, or rose tellin, is one of the commonest of this group, and one that is gathered extensively for the making of shell flowers and other novelties. It is about one inch in length and rosy pink in color, the umbonal area generally darker in hue. The interior has the same shade as the outside. The shell is smooth and delicate, and nearly as high as it is long. The anterior end is rounded, and the posterior end is slightly wedge-shaped. The surface has many very fine concentric lines. The rose petal lives in sand bars in shallow water, and the dainty rose-and-pink valves, looking like so many rose petals, are washed up with each tide. Specimens may be collected from Cape Hatteras south.

TELLINA RADIATA Linne (Rising Sun) p. 28

The rising sun shell is an elongate bivalve, from two to three inches in length. Its color is yellowish white, with bands of pinkish rose radiating from the beaks. The shell is thin but sturdy, with low beaks that are placed about midway of the shell. Both ends are broadly rounded, the posterior one a little less so than the anterior. The surface is smooth and very highly polished. This species, occurring from South Carolina to the West Indies, is one of the prettiest shells to be found on our shores. The broad rosy rays, which are often as vivid on the inside of the shell as they are on the exterior, coupled with the pelecypod's high gloss, make this species a prime favorite with collectors. There is a variety of this species that lacks the pinkish rays. This form is uniform in color, generally white or pale yellow, and it bears the same high gloss of the rayed variety. This form is known as *Tellina radiata unimaculata* Lamarck.

TELLINA INTERRUPTA Wood p. 28

This species gets to be about two inches long, occasionally a little longer. Its color is creamy white, with crowded streaks of brownish purple. The shell is rather long and thin, flattened, and not polished like the last species. The anterior end is rounded, and the posterior end is rostrate. The beaks are nearly central in position but not very prominent. The surface is sculptured with strong, equidistant, concentric lines. The interior of the shell is polished. This is a gayly colored shell, occurring from North Carolina to Brazil. It lives in moderately shallow water, and buries itself deeper in the mud or sand than most bivalves, so that one needs to dig "way down" in order to get living examples. Scattered valves are usually to be seen on Florida beaches.

TELLINA LÆVIGATA Linne   (Smooth Tellin)   p. 200
This shell is about two inches in length, and its outline is roundish — less elongate than many of its group. The anterior end is regularly rounded and moderately inflated, while the slightly shorter posterior end is bluntly pointed, with a distinct angle running from the beaks to that tip. The shell is glossy inside and outside, and the color is white, with pale orange rays. This fine species lives in Florida and the West Indies.

TELLINA LINTEA Conrad   (Linen Tellin)   p. 124
The linen tellin is just under one inch in length, and it is pure white in color. The shell is thin and rather delicate, the beaks small and pointed. The anterior end is rounded, and the posterior end is sharply sloping, with a slightly truncate tip. The surface is sculptured with finely chiseled concentric lines, upturned on the posterior slope to form a filelike edge. The interior of the shell is polished. This neat little shell, almost as high as it is long, lives in moderately deep water. Single valves are not uncommon on the beaches from Cape Hatteras to the Gulf of Mexico.

TELLINA ALTERNATA Say   (Lined Tellin)   p. 28
This is a larger species, averaging about two inches in length. It is commonly white, but may be pale pink or yellow. The shell is compressed, oblong, and narrowed and angulated at the posterior end. The anterior end is gracefully rounded. The surface is decorated with numerous parallel, impressed, concentric lines, every other line obsolete on the posterior area, which is marked by an angular ridge extending from the beaks to the posterior margin. This beautifully and very regularly lined shell is rather common from North Carolina to the Gulf of Mexico. It lives buried in the sand and gravel close to shore, and single as well as connected valves are usually to be found on many of our southern beaches.

TELLINA ANGULOSA Gmelin   (Angled Tellin)   p. 200
This is a really handsome shell, some two inches in length. Its shape is triangular-oval, the beaks forming the apex and about central in position. The posterior end is a little sharper than the anterior, but both ends are slightly ridged. The surface shows fine but distinct concentric lines, and the color is pinkish rose, commonly darker on the interior. This bivalve is found at the Florida Keys. As with many of the tellins, one frequently finds examples with both valves attached, and spread out like so many mounted butterflies.

MISCELLANEOUS GASTROPODS

All shells approximately one-half natural size.

1. *Busycon canaliculatum* 1 view: **CHANNELED PEAR CONCH**
   p. 205
   Suture line channeled.

2. *Polinices duplicata* 1 view: **LOBED MOON SHELL** p. 134
   Brown shelly lobe covering umbilicus.

3. *Busycon pyrum* 1 view: **FIG SHELL**   p. 206
   Spire short: large aperture.

4. *Busycon caricum* 1 view: **KNOBBED PEAR CONCH** p. 203
   Shoulder usually knobby; orange interior.

5. *Ficus papyratia* 1 view: **PAPER FIG SHELL**   p. 177
   Thin-shelled; no spire, flat on top.

6. *Busycon perversum* 1 view: **LIGHTNING CONCH**   p. 205
   Commonly streaked with violet; left-handed.

7. *Morum oniscus* 1 view: **WOOD LOUSE**   p. 175
   Small but heavy; warty surface.

8. *Colus stimpsoni* 1 view: **STIMPSON'S WHELK**   p. 199
   Velvety green periostracum; chalky white inside.

9. *Bulla occidentalis* 1 view: **BUBBLE SHELL**   p. 225
   Aperture longer than body-whorl; smooth surface.

10. *Neptunea decemcostata* 1 view: **TEN–RIDGED WHELK**
    p. 198
    Ten red-brown revolving ridges.

11. *Pyrene mercatoria* 2 views: **MOTTLED DOVE SHELL**
    p. 190
    Narrow aperture, thickest at center.

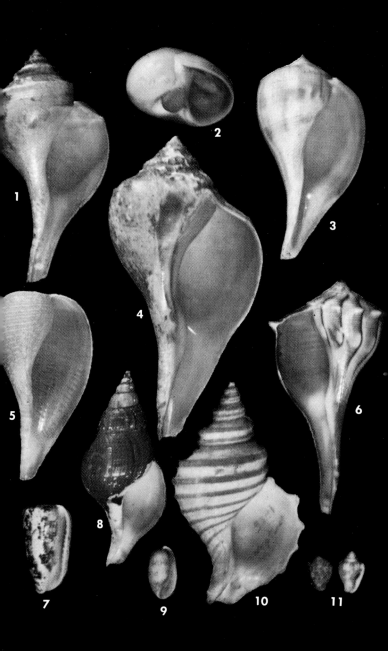

Plate 14   77

MISCELLANEOUS GASTROPODS

All shells approximately one-half natural size.

1. *Murex florifer* 1 view: **BLACK LACE MUREX**   p. 181
   Dark brown or black; aperture and apex often pink.

2. *Murex pomum* 1 view: **APPLE MUREX**   p. 180
   Rough and spiny; mottled brown.

3. *Murex cabritii* 1 view: **SPINY MUREX**   p. 181
   Elongate canal, often with spines.

4. *Distorsio clathrata* 1 view: **WRITHING SHELL**   p. 179
   Constricted aperture; short canal.

5. *Cymatium aquitile* 1 view: **HAIRY TRITON**   p. 177
   Strongly banded varices; checkered columella.

6. *Cymatium cynocephalum*: **RIBBED TRITON**   p. 178
   Broadly rounded revolving ribs.

7. *Charonia tritonis nobilis* 1 view: **TRUMPET SHELL**   p. 179
   Large; highly colored; beaded suture.

8. *Cymatium femorale* 1 view: **ANGULAR TRITON**   p. 178
   Outer lip with large brown knobs.

9. *Cymatium tuberosum* 1 view: **WHITE–MOUTHED TRITON**
                                                            p. 178
   Notch at upper angle of aperture.

10. *Urosalpinx cinerea* 1 view: **OYSTER DRILL**   p. 184
    Small but rugged; purplish interior.

11. *Eupleura caudata* 1 view: **THICK–LIPPED DRILL**   p. 182
    Grayish purple; thickened outer lip.

TELLINA MAGNA Spengler   (Big Tellin)   p. 124

This is the largest member of its group found on our coast, adult shells measuring about four inches in length. The usual color is pinkish yellow and white, but young specimens are often strongly rayed, while old individuals are commonly uniform orange in color. The shell is smooth and polished, with relatively small beaks. The anterior end is long and rounded, while the posterior end is shorter and somewhat angulated. The interior is white, tinged with brownish orange. This handsome pelecypod may be found from the Carolinas to Florida.

TELLINA AURORA Hanley   p. 200

About three-fourths of an inch long, this is an oval shell with the anterior end twice as long as the posterior. Both ends are rounded, the posterior one more acutely. The beaks are fairly prominent. The surface is smooth, and the color is dull white, occasionally pinkish. The interior is shiny, and tinged with yellow. This is an uncommon shell, found offshore in Florida.

Genus *Macoma* Leach 1819

16 Species and varieties

MACOMA BALTHICA Linne   (Baltic Macoma)   p. 97

This common clam is just over one inch in length, and pinkish white in color, with a dull finish. The shell is moderately thin, with rounded outline, the posterior end somewhat constricted. The beaks are rather prominent and nearly central in position. The surface bears numerous very fine concentric lines of growth, and there is a thin, olive-brown periostracum that is usually lacking on the upper parts of the shell. This little clam is abundant in muddy bays and coves, and it commonly travels part way up many creeks and rivers. It thrives all along the Atlantic coast to Cape Hatteras, and occurs in deeper water as far south as Georgia. It is also common in Norway and Sweden.

MACOMA CONSTRICTA Bruguière   (Constricted Macoma)   p. 97

This species is about two and one-half inches long, and white in color, with a thin yellowish periostracum. The shell is moderately inflated, with broadly rounded margins. The posterior end is partially truncate, and notched below, with a feeble fold extending from the beaks to the ventral margin. This bivalve, with its pinched-out posterior end, may be found from Cape Hatteras south. It prefers to live in moderately deep water, but single valves can often be picked up on the beach.

MACOMA BREVIFRONS Say   p. 200

This is a thin and fragile shell, looking very much like one of the tellins. Its length is fully two inches, and its shape is elongate-oval, the beaks a little closer to one end. The anterior end is gracefully rounded, with the posterior end shorter and somewhat sloping. The color is pinkish white, tinged with yellow on the umbones, with the interior commonly darker pink. This species occurs from New Jersey to South America.

MACOMA TENTA Say   p. 97
This is a small, thin, and delicate bivalve. Its length is slightly less than one-half inch. The anterior end is long and broadly rounded, and the posterior end is short and abruptly sloping. The surface bears tiny but sharp lines of growth. The color is pinkish white, but the surface is iridescent, and reflects a whole rainbow of hues. This species lives close to shore, and valves may be found on sandy beaches from Prince Edward Island to the Gulf of Mexico.

Genus *Strigilla* Turton 1822

4 Species

STRIGILLA CARNARIA Linne   (Rosy Carnaria)   p. 28
This little bivalve is slightly under one inch in length. It is pale rose in color, deepest over the umbones, with the interior rich rosy pink. The shell is moderately solid, rather orbicular in outline, and fairly well inflated. The beaks are somewhat elevated. The surface is smooth, with a sculpture of extremely fine radial lines which become oblique and wavy over the posterior area. This colorful shell may be found in quiet bays and inlets from North Carolina to Brazil. It abounds in sand just offshore, and empty valves can generally be gathered by the basketful at the tide marks. The rosy color makes this species an attractive shell for decorative purposes.

Genus *Tellidora* Morch 1856

1 Species

TELLIDORA CRISTATA Recluz   (Saw-tooth)   p. 124
This is an odd little clam, about one inch in length and pure white in color. The shell is compressed, the left valve flatter than the right, and the outline is somewhat triangular. The beaks are central and prominent. The basal margin is broadly rounded. The surface bears faint concentric ridges which form teeth on the lateral margins, giving the shell a saw-toothed appearance. This dainty little shell is not likely to be confused with any other. Its station in life is in the muds and sands close to shore, where it may be collected at low tide. It occurs on the Florida coast of the Gulf of Mexico, and in the West Indies.

Genus *Apolymetis* Salisbury 1929

1 Species

APOLYMETIS INTASTRIATA Say   p. 200
This is a rather large but thin shell, about three inches long.
The posterior end is long and broadly rounded, while the
anterior end is sloping and profoundly folded, so that the shell
has an oddly twisted appearance at its front end. The valves
are well inflated, and pure white in color. This pelecypod likes
to live in brackish situations, and is generally to be found in
the tidal regions of coastal rivers, where it burrows in the mud.
It ranges from southern Florida to the West Indies. This
species is often seen in amateur collections labelled as *Macoma
constricta*, but when the two are viewed together they are
easily separated.

Family Semelidæ

SHELLS ROUNDED-OVAL and but little inflated, with more or less
obscure folds on the posterior ends. They are chiefly confined
to warm seas.

Genus *Semele* Schumacher 1817

5 Species and varieties

SEMELE PROFICUA Pultney   p. 108
This is a creamy-white, sometimes variegated bivalve, about
one and one-half inches in length. The shell is rather thin and
compressed, and rounded or oval in outline. The beaks are
small and turned forward, with a small lunule in front. There
is a sculpture of extremely fine but sharp concentric lines,
and a very thin periostracum. This pelecypod lives buried in
sandy mud, in moderately shallow water, from Virginia to the
West Indies. It possesses a powerful "foot" and is capable of
creeping about over the bottom with considerable agility.

SEMELE PURPURASCENS Gmelin   p. 200
About two inches in length, this is an oval shell, with both
ends rounded. The posterior end is about twice the length of
the anterior. The shell bears fine concentric lines, but the
surface appears quite smooth. The color is pale yellow, more
or less blotched with purplish. Some individuals are uniformly
deep yellow. This species occurs from the Carolinas to the
West Indies.

Genus *Abra* Lamarck 1818

3 Species

ABRA ÆQUALIS Say   p. 60
This is a small, rather plump shell, just under one-half inch in length. Its color is white, often tinged with buff. The shell is rounded but slightly oblique, and it is decorated with minute but numerous concentric wrinkles near the margins, the rest of the surface being relatively smooth and in some cases polished. This little bivalve lives in rather deep water, and is at home from Connecticut to Texas.

Genus *Cumingia* Sowerby 1883

2 Species

CUMINGIA TELLINOIDES Conrad   p. 92
This is a small white clam, about one-half inch in length. The shell is oval-triangular, and quite thin, the anterior end broadly rounded and the posterior end considerably pointed and gaping. The surface is covered by numerous sharp, elevated, concentric lines. This species has a more northern range than many of its family, occurring from Prince Edward Island to Florida.

Family Donacidæ

GENERALLY SMALL, wedge-shaped clams, the posterior end prolonged and acutely rounded, and the anterior end short and sloping. Distributed in all seas that are warm, where they live in the sand close to shore.

Genus *Donax* Linne 1758

7 Species and varieties

DONAX VARIABILIS Say   (Coquina)   p. 12
This small bivalve is known by a variety of names, among them "butterfly shell," "wedge shell," "pompano shell," and "coquina." It is about three quarters of an inch long, and comes in a bewildering variety of colors ranging from pure white to yellow, rose, lavender, pale blue, and deep purple. It is usually decorated with radiating reddish-brown bands, and sometimes with concentric colored lines. Now and then a shell has both, producing a plaid pattern. The shell is triangular, wedge-shaped, and sturdy. The posterior end is prolonged and acutely rounded, while the anterior end is short and obliquely truncated. The

surface is striated with numerous radiating, impressed lines on the anterior end, so fine that they are hardly discernible to the unassisted eye. The inner margins are crenulate, with the inner surface of the shell glossy, and it, too, commonly varies considerably in color. This is a very common little clam on the Florida beaches, and it may be found from North Carolina to Texas. It burrows an inch or so in loose sand at the midwater line where, in favorable situations, individuals may be gathered by the handful with scarcely any intermixture of sand. Despite their diminutive size, they are often so gathered and made into "coquina chowder." Dead shells usually remain in pairs, connected at the hinge and spread out like so many butterflies. The color patterns are almost indefinite, and out of fifty shells it is sometimes difficult to find two that are exactly alike. Suites of this little clam are often used in biology classes to illustrate an extreme in color variation within a single species.

DONAX DENTICULATA Linne   (Toothed Donax) p. 108
This is a sturdier shell, a little more than one inch in length. It may be brown, violet, or yellowish white in color, often with darker rays. The shell is strong and solid, elongate-triangular in outline, with fine radiating lines on the surface. The inner margin is strongly crenulate. This species does not exhibit as great a range of color as *variabilis*, although it, too, is a variable species. It lives in the sand close to shore, and it is sometimes taken in southern Florida, but its real home is in the West Indies.

DONAX FOSSOR Say   p. 200
This is another member of the group, ranging north much further than the others. Its length is about an inch, and the shell is rather thick and solid, with the inner margins crenulate. The surface is decorated with very fine radiating lines, and the color is grayish with whitish or bluish rays. This little fellow lives as far north as the south shore of Long Island. It lives just beneath the surface of the sand in tidal areas, and is especially abundant along the New Jersey coast.

Genus *Iphigenia* Schumacher 1817

1 Species

IPHIGENIA BRASILIANA Lamarck   p. 200
This is a fairly large and sturdy clam, broadly triangular in shape and attaining a length of about three inches. The beaks are prominent, and about central in position, only a little nearer to the posterior end. The anterior end is sharply rounded, and the posterior end even more sharply rounded, with an angled slope at that end. The color is buffy white, sometimes bluish white, and glossy, but in life there is a rather substantial tan

periostracum. The interior is shiny white, often purplish on the hinge teeth. This pelecypod prefers to live in quiet, brackish waters, and it thrives in inland waters that are subject to tidal influence. Its range is from Florida to Brazil.

Family Sanguinolariidæ

SHELLS somewhat like the tellins, with which they were formerly grouped. The animals have long siphons. Distributed chiefly in warm seas.

Genus *Asaphis* Modeer 1793

1 Species

ASAPHIS DEFLORATA Linne   (Rayed Cockle) pp. 28, 96
Sometimes called the "rayed cockle," this is a fair-sized clam, averaging about two inches in length but occasionally growing larger. The color is variable, ranging from white to yellow and orange and purple, with the majority of specimens tending toward the latter shade. The shell is thin but strong, and rather well inflated. The surface bears numerous radiating lines, most pronounced on the posterior slope. The lines at both ends are crossed by wavy growth lines. This is quite a handsome bivalve, common enough in the West Indies but only occasionally found in Florida. It inhabits shallow water in weedy coves. Museums frequently exhibit groups of these pelecypods to show the range of color variation within a species, although this shell is not as striking for that purpose as the little *Donax* just described. Thousands of shells of the rayed cockle are annually exported from Cuba for making up into shell novelties.

Genus *Tagelus* Gray 1847

2 Species

TAGELUS GIBBUS Spengler   (Stout Razor)   p. 109
The stout razor clam grows to a length of about four inches. The shell is white or yellowish, with a thin, yellowish-brown periostracum, but most of the shells that have been empty for any length of time are a dull, chalky white. The shell is elongate, stout, gaping, and abruptly rounded at both ends. The beaks are blunt, and but little elevated, and are situated just off the middle of the shell. The surface is coarsely wrinkled concentrically. This is a common mud-burrower, living in colonies in the mud flats between tides. Its odd, oblong shape renders it easy to identify. The stout razor occurs from Cape Cod southward.

TAGELUS DIVISUS Spengler   p. 109
This is a delicate little fellow, about one and one-half inches
long. Its color is yellowish gray, faintly rayed with purple.
The shell is thin and fragile, elongate, and has parallel margins.
Both ends are bluntly rounded. The beaks are nearly central,
and the surface is smooth, generally shining, with a thin periostracum. This small razor clam may be found living in the
sand, in colonies, all the way from the shore line out to moderate
depths. Its range is from Cape Cod to Mexico, but specimens
are not very abundant in the north. The interior of the shell
is polished and often purple in color.

Family Solenidæ

MEMBERS of this family are the true razor clams, and they are
so called wherever they occur. The shells are equivalve, usually
greatly elongated, and gape at both ends. They are distributed
in the sandy bottoms of coastal waters in nearly all seas. All are
edible.

Genus *Solen* Linne 1758

1 Species

SOLEN VIRIDIS Say   (Little Green Razor)   p. 109
This species is sometimes called the "short razor." It is only
about two or three inches long when fully grown. Green in color,
the shell is thin, elongate, and somewhat compressed, the hinge
line nearly straight, and the valves open at each end. The
surface is smooth and rather glossy, with very faint concentric
wrinkles. Like all the razor clams, this little fellow burrows
vertically in sandy bars that are more or less exposed at low
tide. Specimens may be distinguished from young individuals
of the common razor clam by their straighter shells. This bivalve, which usually lives in colonies, is found from Rhode
Island to Florida. It is abundant on our southern shores, but
rather uncommon north of New Jersey.

Genus *Ensis* Schumacher 1817

2 Species

ENSIS DIRECTUS Conrad   (Common Razor Clam)

p. 109
The common razor clam is familiar to most seaside visitors, and
it probably has the best popular name of any of our mollusks, for

its shape certainly does suggest one of the old straight razors in common use a generation ago. Six or seven inches in length, and yellowish green or olive green in color, the shell is thin, gaping, greatly elongated, and slightly curved. The sides are parallel, and the ends are squarish. The surface has a glossy periostracum and a long triangular space marked by concentric lines of growth. The razor clam occurs all along the Atlantic coast, living in colonies in the sand and mud near the low-water line. It burrows vertically, and the speed with which it can sink down out of sight in wet sand is astonishing, and it is no easy task to dig a specimen out.

This slender bivalve is also a successful, if somewhat erratic, swimmer. One summer the writer discovered a colony on a sand bar that was exposed at low tide, near Harpswell, Maine. The shells stuck up about two inches above the sand, looking like a field of miniature fence posts. Merely walking over the sands caused them to withdraw from sight, and even when a good grip was secured on an individual it was hard to pull the mollusk out without cutting my hand on its sharp edge. By walking cautiously, however, and thrusting a shovel deep into the sand in back of the clam and giving it a quick flip, I managed to obtain a bucketful of fine specimens. Carrying them back to camp, I thought I would wash them off, so I emptied the bucket in shallow water near shore, intending to rinse the razors the same as one ordinarily does "steamer Clams"; whereupon my pile of razors suddenly "exploded" and went shooting off in all directions like so many skyrockets, and most of them I never did recover!

The clam has a very long foot which it protrudes until it is almost if not quite as long as the shell. It then folds this foot back tightly against the shell and suddenly straightens it out as if it were a steel spring, and the clam is propelled like an arrow for some three or four feet. The operation is quickly repeated, and the mollusk goes zigzagging off at a great rate. Laid out on wet sand, the razor clam is able to assume a vertical position and get completely out of sight in about thirty seconds. Razor clams are eaten in many localities. They are tender if not too large, and have an excellent flavor. The writer prefers them to any other species, especially for that delicacy, "fried clams."

Genus *Siliqua* Muhlfeld 1811

2 Species

SILIQUA COSTATA Say　　(Ribbed Pod)　　p. 109
This is an attractive shell, attaining a length of about two inches. Its color is greenish yellow, tinged with purple. The shell is oval-elliptical, thin and fragile, moderately elongate,

and broadly rounded at both ends. The beaks are very low, and situated at the anterior third of the shell. The inside is strengthened by a prominent vertical rib which extends from the beaks, bending slightly backwards, and, expanding, loses itself about halfway across the valve. There is a thin periostracum, and the surface of the shell is smooth, polished, and often iridescent. This is a neat shell, found from the Gulf of St. Lawrence to North Carolina. Its home is in the sands close to shore, and small examples are often taken from the stomachs of fishes.

Genus *Psammosolen* Risso 1826

2 Species

PSAMMOSOLEN CUMINGIANUS Dunker p. 125
This is a queer little pelecypod, not often seen in collections. Its length is about one and one-half inches, and its color is white. Its shape is much like a *Tagelus*, or stout razor clam, with the beaks closer to the anterior end. Both ends are broadly and regularly rounded. The oddity about this shell is its sculpture, consisting of a number of prominent concentric rings, on which is superimposed a series of distinct lines that are oblique on the anterior portion of the valves and sharply vertical on the posterior part. This mollusk lives in rather deep water, and sometimes at great depths, specimens having been taken at more than one hundred fathoms. It occurs from North Carolina to Texas.

Family Mactridæ

SHELLS EQUIVALVE, usually gaping a little at the ends. The hinge has a large, spoon-shaped cavity for an inner cartilaginous ligament. Native to all seas, they live at moderate depths, commonly in the surf.

Genus *Mactra* Linne 1767

1 Species

MACTRA FRAGILIS Gmelin (Fragile Surf Clam) p. 97
The fragile surf clam is about three inches in length, and white or creamy white in color. The shell is thin and moderately delicate, and rather oval in outline, sculptured by very close concentric lines. The beaks are almost central. The anterior end is rounded, while the posterior end is made rostrate by a distinct radiating ridge. The thin periostracum is pale yellow. In spite of its fragile appearance, this pelecypod prefers to live

out in the pounding surf, burrowing through the sand or mud just offshore from North Carolina to Brazil. Perfect dead shells are not easy to find, unless recently cast up on the beach, although the species is far from rare.

Genus *Spisula* Gray 1838

4 Species and varieties

SPISULA SOLIDISSIMA Dillwyn (Surf Clam) p. 97
Called the "surf clam," and "hen clam," this is a large and heavy species, attaining a length of more than six inches. Yellowish white in color, the shell is thick and ponderous, and roughly triangular in outline. The beaks are large and central, with a broad, somewhat flattened area behind them. The hinge is very strong, and there is a large, spoon-shaped cavity within, just under the beaks. The surface is smooth, or slightly wrinkled by growth lines, and there is a thin, olive-brown periostracum. This is the largest bivalve found on our north Atlantic shores. It lives in the surf, where it travels just under the sand, and at high tide the fishermen obtain it by dragging the bottom with sharp sticks. If a stick passes between the open valves of a clam, the mollusk promptly closes upon it and is drawn to the surface. By means of its powerful foot the surf clam is actually able to leap, and thus escape its principal enemies, the giant whelks and the starfishes. Although not as popular as the quahog, this species is regularly eaten, and is considered excellent for "bakes." It occurs from Labrador to North Carolina. In the south its place is taken by a smaller variety known as *Spisula solidissima similis* Say.

Genus *Mulinia* Gray 1837

2 Varieties

MULINIA LATERALIS Say (Little Surf Clam) p. 97
The little surf clam is from one-half to three-fourths of an inch in length, and yellowish white in color. The shell is triangular, smooth and polished, with beaks that are nearly central and inclined forward. The areas before and behind the beaks are broad, flattened, roughly heart-shaped, and bordered by slightly elevated ridges. This species is an important food item for many of our marine fishes, as well as our sea-going ducks. It is sometimes called the "duck clam." It occurs all along the Atlantic coast, from Canada to Mexico. There is a variety of this shell found from North Carolina to Texas that is more triangular, with both margins rather steeply sloping. This form is called *Mulinia lateralis corbuloides* Deshayes.

Genus *Anatina* Schumacher 1817

2 Species

ANATINA CANALICULATA Say   (Channeled Duck)

p. 12

The channeled duck, or hat shell, is about three inches in length and pure white in color. The shell is oval-orbicular, gaping, and very thin and fragile. The posterior slope is narrowed, and the beaks are high, inflated, and directed backward. There is a sculpture of evenly spaced, rounded, concentric grooves. Inside, on the hinge line, is a noticeable spoonlike cavity. This is a thin, delicate, and beautiful pure white bivalve, occurring on the Atlantic coast from Maryland to Brazil, and in the Gulf of Mexico. It lives in sandy mud in shallow water and is commonly washed ashore. The valves are so thin, however, that one seldom is rewarded by finding a complete specimen, although single valves are not at all uncommon.

ANATINA LINEATA Say   (Lined Duck)   p. 96

This species is about the same size, some three inches long, and it is also pure white in color. The shell is thin and swollen, with high beaks. The anterior end is broadly rounded, while the posterior end is gaping and decorated by a cordlike radiating ridge. There is a thin, yellowish periostracum. This, too, is a delicate shell, living in rather deep water from New Jersey to South America. It is not nearly as abundant as the species just described, but single valves are now and then to be picked up on our southern beaches.

Family Mesodesmatidæ

OVAL or wedge-shaped shells, with very short posterior ends. Hinge with a spoon-shaped cavity in each valve. Lateral teeth with a furrow.

Genus *Mesodesma* Deshayes 1830

2 Species

MESODESMA ARCTATA Conrad   p. 200

The outline is rather wedge-shaped, the very short posterior end forming the base of the wedge. The anterior end is narrowed and regularly rounded. The shell is thick and strong, and the color is brown, with a yellowish periostracum that often reflects a metallic luster. The interior is white, with the muscle impressions strongly delineated. The length is from one and one-half to two inches. This rugged bivalve lives in the sand in moderately shallow water, from the Gulf of St. Lawrence to New Jersey.

Family Myacidæ

SHELLS usually inequivalve and gaping. The left valve contains a spoonlike structure called a chondrophore, which fits into a corresponding groove in the right valve. These clams are distributed in all seas.

Genus *Mya* Linne 1758
3 Species and varieties

MYA ARENARIA Linne   (Long Clam)   p. 109
This is the familiar clam that is always to be found on every quiet shore. Known by such names as "long clam," "soft-shelled clam," "steamer clam," and "long-necked clam," it lives in the muds and gravels between the tides, where it is partially exposed to the air twice a day. The long clam gets to be five or six inches long if it is lucky, but persistent "clamming" has made examples that large rare and hard to find. The shell is moderately thick and gapes at both ends. Dull gray or chalky white in color, the surface is roughened and somewhat wrinkled by lines of growth. An erect, spoonlike tooth (chondrophore) is located under the beak in the left valve. This succulent clam lies buried with just the tip of its siphon at the surface. As one walks over its territory the mollusk's position is revealed by a vertical spurt of water, ejected as the alarmed clam suddenly withdraws its siphon. Although generally regarded as inferior to the quahog, this is a very important food mollusk, and it enjoys a steady popularity in the markets. It is common all along the Atlantic coast, as far north as Greenland.

MYA TRUNCATA Linne   (Truncata Mya)   p. 109
This bivalve is from two to three inches long, and dingy white in color. The shell is rather thick and solid, and oblong in shape. The anterior end is rounded, but the posterior end is abruptly truncated, where it gapes widely, the truncated edges flaring more or less. The beaks are moderately prominent. The surface is roughly wrinkled, and there is a thick, tough, yellowish-brown periostracum, which often extends over the truncated part of the valves. This pelecypod looks very much like the common long clam just described, but it is immediately recognized by the peculiar manner in which the posterior end is chopped off. It is a cold-water species, seldom found south of

Maine, although single valves are occasionally washed up in Massachusetts. It occurs very commonly as a Pleistocene fossil throughout the northern New England shore line.

Family Corbulidæ

SHELLS SMALL but solid, very inequivalve, one valve generally overlapping its mate. Slightly gaping at the anterior end. White in color and usually concentrically ribbed. An upright, conical tooth is present in each valve. These bivalves are distributed in nearly all temperate seas.

Genus *Corbula* Bruguière 1797

10 Species

CORBULA CONTRACTA Say   (Basket Shell)   p. 97
The little basket clam is only about one-half inch in length, and dull white in color, with a thin, brownish periostracum. The shell is solid and convex, the anterior end rounded and the posterior end somewhat pointed. An angular ridge runs from the beaks to the posterior basal margin, giving that portion of the shell a distinct slope. The surface is sculptured with regular, smooth, concentric ribs. The basket clam lives in the sands and muds of shallow water, and may be found from Cape Cod to the West Indies. The regular concentric ribs are characteristic of this species, and make it easy to recognize.

CORBULA SWIFTIANA Adams   (Swift's Basket Shell)
p. 97
This is a slightly smaller species, averaging less than one-half inch long. The white shell is somewhat triangular, the posterior ridge quite prominent. The concentric ribs are feeble or lacking in some instances, and there may be a faint trace of revolving lines, especially toward the two ends. This species occurs from Cape Hatteras to South America.

Genus *Paramya* Conrad 1860

1 Species

PARAMYA   SUBOVATA Conrad   p. 97
This is a small pelecypod, about one-half inch long. Its color is dull white or yellowish white. The shell is rather squarish in outline, the anterior end rounded and the posterior end enlarged, and bluntly rounded. The beaks are moderately prominent. This little fellow lives in fairly shallow water, from North Carolina to the west coast of Florida.

Family Saxicavidæ

SHELLS usually elongate and inequivalve. Members of this family commonly bore into sponge, coral, and limestone. The surface of the shell is irregular and rough. They range from the Arctic to the Tropics.

Genus *Saxicava* Bellevue 1802

3 Species

SAXICAVA ARCTICA Linne   (Arctic Rock-borer)   p. 124
This is a rough and unattractive bivalve, about one and one-half inches long and dingy white in color. The shell is oblong-oval, and coarse and irregular in shape. The beaks are rather prominent, and from them run two faint ridges to the posterior margin. Both ends are rounded, with the posterior end nearly three times as long as the anterior. The surface is coarsely marked with lines of growth and is irregularly undulated. This is a boring clam, sometimes living in hard-packed clay, but frequently boring into limestone. It occasionally does considerable damage by excavating its burrows in the cement work of breakers or embankments. Each individual, after boring its home, attaches itself to the walls of its burrow by a byssal cord and remains fixed for life. This is a cold-water clam, and single valves are not uncommon on northern beaches. Like *Mya truncata* that was recently discussed, this shell is quite common in fossil deposits of Pleistocene age, throughout the northeast.

Genus *Panope* Menard 1807

1 Species

PANOPE BITRUNCATA Conrad   p. 125
This is a large and stout clam, five or six inches in length, and pure white in color. The shell is well inflated, with prominent beaks, regularly rounded before and rather sharply truncated behind, where it gapes widely. The surface is quite smooth, marked only by the rather obscure growth lines. This fine large shell occurs in the sands and muds of shallow water, from the Carolinas south.

Genus *Panomya* Gray 1857

1 Species

PANOMYA ARCTICA Lamarck   p. 200
This is a rough and sturdy shell, rather squarish in outline, attaining a length of about three inches. The posterior end is

THE ASTARTES AND SOME OTHERS

1. *Lyonsia hyalina* × 1, 2 views: **GLOSSY LYONSIA**   p. 47
   Thin-shelled; iridescent; inflated.

2. *Pandora gouldiana* × 1, 3 views: **GOULD'S PANDORA**
                                                         p. 46
   Extremely thin; pearly inside.

3. *Pandora trilineata* × 1, 2 views: **THREE–LINED PANDORA**
                                                         p. 46
   Like 2, but more elongate.

4. *Cuspidaria costellata* × 2, 1 view: **COSTELLATE DIPPER**
                                                         p. 49
   Tiny, with dipperlike "handle."

5. *Pandora glacialis* × 1, 1 view   p. 47
   Like 2, but larger.

6. *Cumingia tellinoides* × 1, 2 views   p. 81
   See text.

7. *Astarte striata* × 1, 3 views: **STRIATE ASTARTE**   p. 53
   Small; brown; concentric furrows.

8. *Astarte castanea* × 1, 2 views: **CHESTNUT ASTARTE** p. 52
   Brown; surface fairly smooth.

9. *Astarte undata* × 1, 2 views: **WAVED ASTARTE**   p. 52
   Brown; strong concentric furrows.

10. *Astarte subæquilatera* × 1, 2 views: **LENTIL ASTARTE**
                                                         p. 53
    Brown; furrows less broad than 9.

11. *Venericardia borealis* × 1, 2 views: **NORTHERN CARDITA**
                                                         p. 54
    Small but solid; strong radiating ribs.

12. *Polymesoda floridana* × 1, 2 views: **FLORIDA POLYME–
    SODA**   p. 51
    Smooth; commonly purplish.

13. *Cardita gracilis* × 1, 2 views: **GRACEFUL CARDITA** p. 54
    Elongate; dark interior.

14. *Polymesoda carolinensis* × 1, 2 views: **CAROLINA POLY–
    MESODA**   p. 51
    Greenish, often eroded.

Plate 16   93

THE LUCINES

1. *Lævicardium mortoni* × 1, 2 views: **MORTON'S COCKLE**
   p. 65
   Small, glossy.

2. *Loripinus schrammi* × 1, 1 view: **SCHRAMM'S LORIPINUS**
   p. 59
   Large, inflated; dull white.   •

3. *Lucina floridana* × 1, 2 views: **FLORIDA LUCINE**   p. 57
   Thin: fold not prominent.

4. *Codakia orbiculata* × 1, 1 view: **LITTLE WHITE LUCINE**
   p. 58
   Small; fine sculpture.

5. *Lucina pennsylvanica* × 1, 2 views: **PENNSYLVANIA LU-CINE**   p. 56
   Nearly circular; sharp concentric ridges.

6. *Codakia orbicularis* × 1, 2 views: **GREAT WHITE LUCINE**
   p. 58
   Large; white with pinkish border.

7. *Lucina jamaicensis* × ½, 2 views: **JAMAICA LUCINE**   p. 56
   Yellowish white; widely spaced ridges.

8. *Trachycardium egmontianum* × 1, 2 views: **CHINA COCKLE**
   p. 59
   Interior reddish purple.

squarely cut off, and the anterior end slopes to a rounded point and then rounds to the basal margin, which is straight. The color is chalky white, and there is a dark, almost black periostracum that peels easily from the shell, so that even living examples are rarely completely covered. The posterior end gapes widely, and the shell fails to completely encase the animal within. This bivalve burrows in the mud of gravelly shores between tides. The distribution is circumpolar, and on our coast it gets down as far as Maine. It used to be listed as *Mya norvegica* Spengler.

Genus *Cyrtodaria* Daudin 1799

1 Species

CYRTODARIA SILIQUA Spengler   p. 200
This is a thick and heavy shell, about three inches in length. Its shape is elongate and oval, with the beaks closer to the anterior end. Both ends are rounded, and gape widely. The surface bears concentric grooves, and there is a thick and horny, glossy black periostracum that frequently projects beyond the valves. The interior is white. The beaks are very low, and are usually eroded. The animal is very large, and the valves never completely cover the fleshy parts. This bivalve occurs from Newfoundland to Cape Cod.

Family Gastrochænidæ

BURROWING or boring mollusks, living in coral, limestone, or dead shells of other mollusks. The shells are equivalve and gape considerably.

Genus *Gastrochæna* Spengler 1783

4 Species

GASTROCHÆNA OVATA Sowerby   p. 200
This is a small, oddly shaped boring clam about one inch or less in length. The oval shell gapes widely, and is somewhat twisted in appearance. The beaks are set close to the anterior end, and that end slopes sharply back to the basal margin. The color is yellowish gray, and the valves bear very fine concentric lines. This species excavates burrows in other shells, especially large examples of *Spondylus*. Its range is from South Carolina to the West Indies and Mexico.

GASTROCHÆNA ROSTRATA Spengler   p. 124
This is a small, yellowish-white bivalve, about one inch in length. The shell is thin but sturdy, elongate, squarish, slightly

twisted, and widely gaping. The anterior end is rounded, and the posterior end is squarely truncated. There is an elevated, transversely ribbed, radiating area extending from the beaks to the posterior margin. This boring clam makes its home in coral and limestone. It may be found in southern Florida and the West Indies. It is abundant in the latter islands, but rare in this country.

Family Pholadidæ

THESE ARE boring clams, capable of penetrating wood, coral, and moderately hard rocks. The shells are white, thin, and brittle, generally elongate, and narrowed toward the posterior end. There is a sharp, abrading sculpture on the anterior end. They gape at both ends. Distributed in all seas.

Genus *Pholas* Linne 1758

1 Species

PHOLAS CAMPECHIENSIS Gmelin   (Wing Shell)

p. 124

The wing shell is a slender, pure white bivalve, about three inches in length. The shell is greatly elongated, thin, and brittle, gaping at both ends. It is rayed all over with rather distant ribs, those of the anterior end sharp and rasplike. The hinge plate is reflected over the umbones. This graceful shell closely resembles the famous "angel's wing," to be discussed next, but it is smaller and slimmer, with a sculpture that is much finer. It occurs from Cape Hatteras to South America, and lives in colonies in vertical burrows in the sand close to shore.

Genus *Barnea* Risso 1826

3 Species

BARNEA COSTATA Linne   (Angel's Wing)   p. 124

This is the angel's wing, and a single glance at the snowy-white, graceful shell is enough to convince one that the bivalve has been well named. Six to seven inches in length, the shell is thin and brittle, rounded before, and narrowed and prolonged behind. It is sculptured with strong radiating ribs, about one-half inch apart on the basal margin, becoming abruptly closer as they approach the umbones. Coarse lines of growth rise over the ribs in an undulating manner. The valves gape widely, touch-

THE COCKLES AND SOME OTHERS

1. *Cerastoderma pinnulatum* × 1, 2 views: **LITTLE COCKLE**
   p. 63
   Small; radiating ribs.

2. *Asaphis deflorata* × 1, 1 view: **RAYED COCKLE**   p. 83
   Fine ribs, inflated; pinkish.

3. *Clinocardium ciliatum* × 1, 2 views: **ICELAND COCKLE**
   p. 63
   Large and inflated; strong ribs.

4. *Cyprina islandica* × ⅔, 1 view: **BLACK CLAM**   p. 50
   Heavy; black or deep brown.

5. *Antigona listeri* × ⅔, 2 views: **LISTER'S VENUS**   p. 68
   Strong and solid; close network sculpture.

6. *Pitar morrhuana* × 1, 2 views   p. 68
   Dull; rusty.

7. *Anatina lineata* × 1, 1 view: **LINED DUCK**   p. 88
   Thin-shelled; swollen; white.

8. *Dosinia discus* × 1, 2 views: **DISK SHELL**   p. 66
   Orbicular; close concentric ridges.

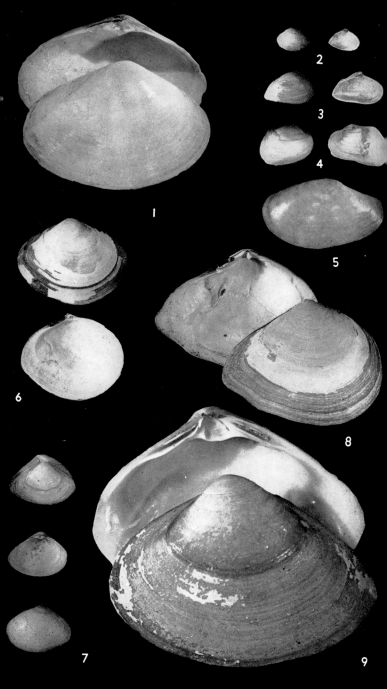

Plate 18   97

THE MACTRAS AND THE BASKET CLAMS

1. *Mactra fragilis* × 1, 2 views: **FRAGILE SURF CLAM**   p. 86
   Radiating ridge on posterior end.

2. *Corbula swiftiana* × 1, 2 views: **SWIFT'S BASKET SHELL**
   p. 90

   See text.

3. *Corbula contracta* × 1, 2 views: **BASKET SHELL**   p. 90
   Small, concentric ribs.

4. *Paramya subovata* × 1, 2 views   p. 90
   See text.

5. *Macoma tenta* × 2, 1 view   p. 79
   Delicate; smooth.

6. *Macoma balthica* × 1, 2 views: **BALTIC MACOMA**   p. 78
   Roundish; dull pinkish white.

7. *Mulinia lateralis* × 1, 3 views: **LITTLE SURF CLAM** p. 87
   Small; triangular; glossy.

8. *Macoma constricta* × 1, 1 view: **CONSTRICTED MACOMA**
   p. 78

   Large: swollen; irregular.

9. *Spisula solidissima* × ⅔, 2 views: **SURF CLAM**   p. 87
   Large; spoonlike cavity under beak.

ing only at a point near the top. The angel's wing has been very dear to the heart of the shell collector since away back. It is a burrowing species, found in colonies several inches deep in sandy mud. It occurs from Cape Cod to the West Indies. Very common in Florida, it is rare north of Virginia. In Cuba the angel's wing is a staple article of food.

BARNEA TRUNCATA Say (Truncated Borer) p. 124
This shell is about two inches in length, and pure white in color. The valves are thin and fragile, somewhat oblong, with the posterior end broadly truncated at the tip. The surface is transversely and longitudinally wrinkled, and is studded, particularly on the anterior end, with small erect scales. The valves gape widely, and there is a small shelly plate situated between the valves, just in front of the beaks. This bivalve burrows in mud or peat banks that are exposed at low tide. The shell is so fragile that it is difficult to dig a specimen free without crushing it. The best way to collect perfect examples of this clam is to dig out a large block of mud from some locality where they are suspected of living and then place the block in the nearest tide pool, where it may be slowly dissolved, thus washing the earth from the clam rather than attempting to pry the clam out of the earth. The truncated borer occurs from Massachusetts to Florida.

Genus *Zirfæa* Gray 1847

1 Species

ZIRFÆA CRISPATA Linne (Great Piddock) p. 124
This bivalve is two or three inches long, and grayish white in color, sometimes a little rusty. The shell is rather oblong, rounded posteriorly, and narrowed and somewhat pointed anteriorly. It is widely gaping at both ends. The surface bears numerous coarse, concentric wrinkle-like ridges which become lamellar on the anterior half of the shell. The laminæ are divided into nearly equal portions by a broad channel running from the beaks to the middle of the basal margin. When living, a membracous expansion covers the upper part of the shell. This, too, is a burrowing clam, showing a marked preference for stiff clay. After a brief free-swimming period the young piddock bores into a clay bank, the burrow being constructed by constantly turning the shell about in close quarters, so that the filelike anterior surface wears away the clay. This is a cold-water pelecypod, found sparingly south to New England. It also occurs in Europe, and a closely related species is found on our west coast.

Genus *Martesia* Leach 1825

5 Species

MARTESIA STRIATA Linne   (Wood Piddock)   p. 125
The wood piddock is about one inch long and grayish white in color. The shell is wedge shaped, the anterior margin nearly closed, the posterior margin with a rounded lip and coarse, filelike ridges. The surface is transversely striated with elevated, minutely crenulated lines. The posterior end tapers by nearly straight edges to a rounded tip. The wood piddock may be collected by searching in old, waterlogged timbers cast up on the beach. It penetrates an inch or more in the wood, commonly working against the grain. The boring is done with the long, elastic foot, and once safely ensconced in its wooden burrow, the piddock is secure from the many enemies that constantly threaten the majority of our bivalves. This species occurs from South Carolina to the West Indies. *Martesia pusilla* Linne and *Martesia cuneiformis* Say are similar species that may be found on almost any of our shores, since they live in timbers that may float in from anywhere.

Family Teredidæ

THESE ARE the ship worms. The anterior part is covered by a bivalve shell, but the posterior part is encased in a long shelly tube. The shell is white, very much inflated, and gaping, and is often decorated with closely set lines. There are a pair of calcareous structures called pallets at the posterior end which are used to close the tube, and the distinctive characters of these pallets are factors in determining the different species.

Genus *Teredo* Linne 1758

15 Species

TEREDO NAVALIS Linne   (Common Ship Worm)   p. 125
The ship worm is a curious little pelecypod, possessing a small, globular, bivalve shell, and is about one-fourth of an inch long so far as this vestigial shell is concerned, but most of the animal lives in a shelly tube that may be as long as six inches. The ship worm lives in wood, and is a great destroyer of timbers, both in ships and in wharves. Infested timbers are so honeycombed by the elongate galleries that they eventually disintegrate and crumble away. Painting or soaking the wood with creosote appears to discourage the mollusk to some extent. Although this little clam costs the shipping industry vast sums

of money yearly, its record is not entirely bad, for it must be said that it performs a valuable service as a scavenger, ridding our harbors of floating and sunken timbers, hulks of derelicts, and other hazards to navigation. This particular species is distributed from the Arctic Seas to the Tropics, and there are few beaches where one cannot collect bits of wood containing their erratic burrows.

Besides *Teredo*, this family contains three other genera of mollusks commonly referred to as "ship worms." These three are *Bankia* Gray 1840, *Xylotrya* Leach 1830, and *Xylophaga* Turton 1822 (p. 200). They all have the same habits of burrowing with the grain of the wood, and their specific identification is generally a task for the experts.

Part Two

The Gastropods

Family Acmæidæ

SHELLS CONICAL somewhat depressed, oval and open at the base. There is no opening at the apex. Not spiral at any stage of growth. Mollusks of the shore region, adhering to rocks and grasses.

Genus *Acmæa* Eschscholtz 1830

8 Species and varieties

ACMÆA TESTUDINALIS Muller (Tortoise-shell Limpet)
p. 128

This univalve is known as the "tortoise-shell limpet." It averages about an inch in length, although some individuals close to two inches long have been collected. The shell is conical, oblong-oval, and only moderately arched, with the apex behind the middle and turning slightly toward the short end. It is not spiral at any stage of growth. The color is bluish white, checkered with dark brown marks radiating from the summit. The inside is dark glossy brown, with a checkered gray and brown border, the two separated by a paler band.

These limpets will be found adhering to the sides of rocks in shallow water and in tide pools, and so tightly do they cling that it is next to impossible to remove one without breaking its shell. Slip a thin-bladed knife under a specimen, however, and it will roll off in your hand uninjured.

The tortoise-shell limpet is a cold-water gastropod, found from Labrador to Connecticut. The largest specimens are found in the vicinity of Eastport, Maine.

ACMÆA TESTUDINALIS ALVEUS Conrad (Eel-grass Limpet)
p. 204

This little fellow is a variety of the shell just described. It is small, thin, and well arched, with the sides parallel and often compressed. Its length is seldom more than one-half inch. The surface is boldly checked with yellowish or whitish dots which are often plainer on the inside of the shell than on the outside. This limpet lives almost exclusively on the narrow leaves of the eel-grass, all along the New England coast.

ACMÆA CANDEANA Orbigny (Southern Limpet) p. 128
This species lives in southern Florida and the West Indies. The shell is about one inch long, and gray or buff in color, with

MISCELLANEOUS GASTROPODS

All shells approximately one-half natural size.

1. *Conus regius* 1 view: **CLOUDY CONE**   p. 217
   Heavily marbled; cone-shaped.

2. *Conus floridanus* 1 view: **FLORIDA CONE**   p. 217
   Pointed spire; band on last whorl.

3. *Conus citrinus* 2 views: **MOUSE CONE**   p. 217
   Small, stubby; blunt spire.

4. *Conus spurius atlanticus* 1 view: **ALPHABET CONE** p. 216
   Large; decorated with "Chinese letters."

5. *Cancellaria reticulata* 1 view: **NUTMEG SHELL**   p. 222
   Orange bands; cancellated surface.

6. *Epitonium eburneum* 1 view: **IVORY WENTLETRAP** p. 122
   Elongate; strong vertical ribs; circular aperture.

7. *Strombus gallus* 1 view: **COCK STROMB**   p. 166
   Greatly expanded, winglike outer lip.

8. *Janthina janthina* 1 view: **VIOLET SNAIL**   p. 123
   Thin-shelled; pale and dark lavender.

9. *Strombus pugilis* 1 view: **FIGHTING CONCH**   p. 164
   Strong and solid; shoulders usually knobby; notch in lower
   outer lip.

10. *Architectonica granulata* 1 view: **SUNDIAL**   p. 146
    Flatly coiled; deep, crenulate umbilicus.

11. *Strombus raninus* 1 view: **HAWK WING**   p. 165
    Large knobs on body-whorl; white interior.

Plate 20   105

MISCELLANEOUS GASTROPODS

All shells approximately one-half natural size.

1. *Thais deltoidea* 1 view: **BANDED DYE SHELL**   p. 187
   Short and squat; purplish interior.

2. *Purpura patula* 1 view: **WIDE–MOUTHED DYE SHELL**
                                                    p. 185
   Extremely large aperture; orange inner lip.

3. *Fasciolaria distans* 1 view: **BANDED TULIP SHELL** p. 207
   Black threadlike revolving lines.

4. *Terebra dislocata* 1 view: **LITTLE SCREW SHELL**   p. 215
   Beaded line below suture.

5. *Fasciolaria gigantea* 1 view: **HORSE CONCH**   p. 208
   Large and massive; reddish when young.

6. *Scaphella junonia* 1 view: **JUNO'S VOLUTE**   p. 211
   Spiral rows of brown spots.

7. *Fasciolaria tulipa* 1 view: **TULIP SHELL**   p. 207
   Reddish, without the black lines of 3.

8. *Melongena melongena* 1 view: **BROWN CROWN CONCH**
                                                    p. 207
   Yellow bands; spines lacking or few in number.

9. *Thais lapillus* 2 views: **ROCK PURPLE**   p. 186
   Small but stout; outer lip thick.

10. *Melongena corona* 1 view: **CROWNED CONCH**   p. 206
    Shoulders with strong curved spines

narrow black lines radiating from the apex. These lines sometimes combine to form broad rays. The shell is conical and but
little elevated, with the apex behind the middle and turning
slightly toward the short end. Like the tortoise-shell limpet
of the North, this species is to be found in shallow water, clinging
to seaweeds and rocks. A variety of this limpet, *Acmæa candeana
antillarum* Sowerby, is more highly arched, and commonly has
the radiating lines bluish on a pale background.

Family Lepetidæ

SHELL SMALL, oval, conical, and somewhat depressed. Apex nearly
central   Distributed in cold seas.

Genus *Lepeta* Gray 1847

1 Species

LEPETA CÆCA Muller   p. 128
This is a tiny limpet, usually not over one-quarter of an inch
long. The shell is conical but somewhat depressed, and oval,
with numerous minute radiating ribs that are crossed by equally
fine concentric lines. The resulting sculpture, when viewed
under a lens, is a distinct network. The color is white or grayish
white, both inside and outside. This is not a very common shell.
It occurs from Greenland to Massachusetts.

Family Fissurellidæ

SHELLS CONICAL, oval at base. The apex is perforated, or there is
a slit, or notch, in the margin of the shell. The surface is usually
strongly ribbed. Distributed in warm and temperate seas.

Genus *Fissurella* Bruguière 1791

5 Species

FISSURELLA BARBADŒNSIS Gmelin   (Barbado Limpet)
                                                                              p. 128
This species is sometimes called the "Barbado chink." It is
nearly one and one-half inches long, and quite variable in color,
generally some shade of gray, green, or pink, with brown blotches.
The shell is solid, conical, and highly elevated, with a round
or oval orifice at the summit. The surface is rough, with heavy
radiating ribs which project at the margins, so that the shell

has a scalloped edge. The interior usually has alternating rings of dull green and white.

The slit, or "key-hole," distinguishes these snails from the true limpets. When very young, these mollusks have a spiral shell with a marginal slit. Shelly material is added slowly until the margin below the slit is united, and then the spiral is absorbed as the hole enlarges.

This species is very common in the West Indies and may be found occasionally in southern Florida. The mollusks live attached to rocks and wharf pilings, as a rule in situations where the waves are continually breaking over them, and their shells are commonly fairly well covered with marine growths.

FISSURELLA ROSEA Gmelin (Rosy Fissurella) p. 128
This shell is about one inch in length, and it is also rather variable in color. The usual shade is delicate pink with pale rays, but some individuals are deep pink or purple. The shell is conical, moderately thin, and but little elevated. The summit bears a small, usually round opening. The marginal outline of the shell is egg-shaped. The sculpture consists of close radiating lines, unequal in size. This limpet is found in southern Florida. Empty shells are now and then found on the beach, but living specimens are not commonly seen as the mollusk lives well below the low water level.

FISSURELLA NODOSA Born (Knobby Limpet) p. 128
This is a large and rugged species, about one and one-half inches long. Its color may be white, brown, gray, or pinkish. The conical shell is moderately elevated, with an elongate opening at the top, and is sculptured with about twenty radiating ribs bearing nodules concentrically arranged. The base is broadly oval, and the margins are deeply scalloped by the ends of the ribs. This species occurs in the Florida Keys, but it is considered quite rare in this country. It is abundant in the West Indies.

FISSURELLA PUNCTATA Fischer (Rocking-Chair Limpet) p. 204
This is a rather flattish limpet, averaging about three-fourths of an inch in length. The base is oval, and the front and back margins are raised noticeably. When placed on a perfectly level surface the shell rests only upon the middle of each side. The apex has a rather long and narrow opening, with a small notch on each side at the center, so that a sort of cross is formed. The sculpture consists of numerous radiating ribs, and the color is grayish or yellowish, usually with a few reddish rays near the perforation. Clench has stated that many specimens appear to have red spots which are in reality foraminifera which cement

THE VENUS CLAMS

1. *Anomalocardia cuneimeris* × 1, 2 views: **POINTED VENUS**
   p. 71
   Small; posterior end prolonged.

2. *Venus mercenaria* × ½, 2 views: **QUAHOG** p. 70
   Roudish; purple border on inside.

3. *Venus mercenaria notata* × ⅔, 2 views p. 70
   Smaller, with dark zigzag lines.

4. *Gemma gemma* × 6, 1 view: **GEM SHELL** p. 72
   Tiny; pale lavender.

5. *Semele proficua* × 1, 2 views p. 80
   See text.

6. *Serripes grœnlandicus* × 1, 2 views: **GREENLAND COCKLE**
   p. 65
   Large but thin; grayish.

7. *Thracia conradi* × 1, 2 views: **CONRAD'S THRACIA** p. 43
   Perforation in one beak.

8. *Donax denticulata* × 1, 2 views: **TOOTHED DONAX** p. 82
   Wedge-shaped; crenulate margin.

Plate 22   109

THE RAZOR CLAMS AND OTHERS

1. *Anomalocardia brasiliana* × 1, 1 view: **LITTLE STRIPED VENUS**   p. 71
   Prolonged posterior end.

2. *Tagelus divisus* × 1, 2 views   p. 84
   Shiny; often rayed.

3. *Tagelus gibbus* × 1, 1 view: **STOUT RAZOR**   p. 83
   Dull; much larger than 2.

4. *Solen viridis* × 1, 2 views: **LITTLE GREEN RAZOR**   p. 84
   Nearly straight; green.

5. *Ensis directus* × 1, 2 views: **COMMON RAZOR CLAM**
   p. 84
   Curved; deep green.

6. *Siliqua costata* × 1, 2 views: **RIBBED POD**   p. 85
   Heavy vertical rib inside shell.

7. *Mya arenaria* × 1, 1 view: **LONG CLAM**   p. 89
   Dull, chalky; fairly thin-shelled.

8. *Mya truncata* × 1, 2 views: **TRUNCATE MYA**   p. 89
   Posterior end chopped off.

themselves to this shell. It may be found from North Carolina
to the West Indies, and for many years it has gone under the
name of *Fissurella pustula* Lamarck, but that name is now restricted to an eastern Atlantic form.

Genus *Lucapina* Gray 1857

4 Species

LUCAPINA SOWERBII Sowerby   (Cancellated Limpet)

p. 128

This species is about one inch in length, and buffy or grayish
white in color. The shell is conical, somewhat depressed, and
the margin is oval. The apex, or summit, with a round orifice,
is slightly ahead of the middle of the shell. The surface is ornamented with radiating ribs, alternating in size and cancellated
by strong, regular, concentric ridges. The margins of the shell
are finely crenulate.

In life, the shell of this snail is almost completely imbedded
in the mantle. There is often a bluish stain around the orifice,
on the inside. Like the other gastropods of this type, the cancellated limpet may be found clinging to the sides of stones
in shallow water. Its range is from Florida to Brazil.

Genus *Lucapinella* Pilsbry 1890

2 Species

LUCAPINELLA LIMATULA Reeve   p. 204

This is a small oval shell, seldom exceeding three-fourths of an
inch in length. It is only moderately arched, with the perforation at the center of the shell. This orifice, relatively large, is
rather triangular in shape. The sculpture consists of alternating
large and small ribs at the anterior end, with two or three small
ribs between the larger ones on the posterior end. Concentric
wrinkles make these radiating ribs scaly. The color may be
white, pinkish, or brown, with the inner surface white. Occurring
from North Carolina to the West Indies, this is a rather rare shell
in this country. It lives in several fathoms of water.

Genus *Diodora* Gray 1821

8 Species and varieties

DIODORA LISTERI Orbigny   (Key-hole Limpet)   p. 128

This is commonly called the "key-hole limpet," although that
name would apply equally well to several other species. The
length is about an inch and a half, and the color is white or
buff, sometimes decorated with black lines. The shell is solid,
oval, and highly elevated. The apex is slightly in front of the

middle of the shell and has a slitlike opening, frequently stained blue-black on the outside. The surface has alternating large and small ribs, crossed by strong cordlike lines. The margins are scalloped by the ends of the ribs. The keyhole limpet is found in shallow water, clinging to rocks, coral, or wharf piles. It occurs from Florida south to British Guiana and is also found in Bermuda.

DIODORA CAYENENSIS Lamarck   (Little Key-hole Limpet)   p. 128

This species is about one inch in length, and like many of the limpets it is quite variable in color. The common shade is white, or grayish white, but it may be buff, pink, or deep gray, with or without dark radiating lines. The shell is moderately solid and highly elevated, with an elongate slit at the summit. The surface is sculptured with numerous distinct radiating lines, every fourth one larger, and wrinkled by concentric growth lines. The interior is white or bluish gray, often polished. The margins are finely crenulate.

This little limpet, which will be found listed in many books under its old name of *Diodora alternata* Say, is much like the last species, *listeri*, but it is smaller, and less coarsely sculptured. It may be found quite plentifully from Chesapeake Bay to Florida, generally upon stones, seaweeds, or shells, in shallow water.

Family Trochidæ

HERBIVOROUS snails, widely distributed in warm seas, living among seaweeds in shallow water, and upon wave-washed rocks. The shells are composed largely of iridescent nacre, although the pearly luster may be concealed by a periostracum during the animal's life. They are varied in shape, but are commonly pyramidal. The operculum is thin and spiral.

Genus *Tegula* Lesson 1832

4 Species

TEGULA FASCIATA Born   p. 129

This shell is about three-fourths of an inch high, and pink or grayish pink in color, rather heavily mottled and marked with brown or black. There is usually a paler band on the body whorl. The shell is roundly pyramidal, consisting of four or five whorls, with a smooth, sometimes shiny surface. The aperture is oval, feebly toothed within, and there are two teeth at the base of the columella. There is a white callus that may extend partially over the umbilicus. The operculum is thin and corneous. This is a very pretty little shell, quite colorful when freshly taken, and very pearly when the outer surface is removed and the shell polished. It is found on plants in shallow water, in southern Florida.

TEGULA EXCAVATA Lamarck   p. 201
This snail is about three-fourths of an inch tall, and its shape is toplike. There are four or five flattish whorls, with no shoulders at all, but with sharply delineated sutures. The base is concave, the aperture rather small, and there is a distinct umbilicus. The color is purplish brown or black, but most specimens are eroded in places to show the pearly layer. The aperture is often greenish. This species is found in southern Florida.

Genus *Livona* Gray 1842

1 Species

LIVONA PICA Linne   (Magpie Shell)   p. 29
The magpie shell is an attractive and striking species, three to five inches in diameter. Its color is black, heavily splashed with zigzag markings of white. The shell is large, solid, and top-shaped, with five or six whorls and a moderately sharp apex. The surface is irregular, the umbilicus deep, and the aperture roundish. The operculum is leathery, and decorated with numerous whorls.

This large and showy species is found as a dead shell in southern Florida, and alive in the West Indies, where it lives on weedy bottoms in shallow water. It is frequently seen in old Indian shell heaps along the Florida coast, suggesting that the species may have been abundant on our shores in the past, but there are only one or two questionable records of live examples occurring there in the last decade. It occurs quite commonly as a Pleistocene fossil in Florida and Bermuda. The shell is very pearly, and when the outer layer has been removed, it can be given a high polish.

Genus *Calliostoma* Swainson 1840

28 Species and varieties

CALLIOSTOMA JUJUBINUM Gmelin   (Mottled Top
Shell)   p. 129
The mottled top shell is about one inch high, and is brownish in color, with gray and white streaks. The shell is robust and conical, with the apex acutely pointed. There are eight to ten whorls, with indistinct sutures. The surface is sculptured with

many revolving cords, those on the shoulders broken up into beads. The umbilicus is narrow and funnel-like, and the operculum is thin and horny. This species is well named, as its shape closely resembles an inverted top. It occurs from Cape Hatteras to the West Indies. It may be taken by dredging in fifteen to thirty feet of water, particularly on weedy bottoms. Empty shells are fairly common on the beaches throughout its range.

CALLIOSTOMA EUGLYPTUM Adams (Florida Top Shell) p. 129

The Florida top shell is about one inch high. Its color is dingy white, boldly mottled with reds and browns. The shell is strong and solid, top-shaped, and highly elevated, with the apex sharply pointed and the base flattened. There are five or six whorls, with rounded shoulders, the sutures rather indistinct. The ornamentation consists of revolving rows of beaded ridges. The columella is oblique, and thickened at the base. This is a real colorful shell when taken fresh. It occurs on seaweeds in moderately shallow water, from Cape Hatteras to Mexico. It is easily distinguished from the last species, *jujubinum*, by its gently rounded shoulders.

CALLIOSTOMA BAIRDII Verrill & Smith (Baird's Top Shell) p. 201

This is a fine shell, the largest and handsomest of its genus. Fully an inch in height and a little more in diameter, it tapers regularly to form a flattish base and a sharply pointed apex. There are about six whorls, with no shoulders, and the sutures are indistinct. Each volution is decorated with revolving rows of small beads, with one row near the suture larger than the others. The color is yellowish brown, more or less spotted with squarish red dots. Like the others of its group, this shell has a pearly layer under its pebbly exterior. It lives in rather deep water, from Massachusetts to the Florida Keys.

CALLIOSTOMA OCCIDENTALIS Mighels & Adams (Pearly Top Shell) p. 129

The pearly top shell is about one-half inch in diameter, and creamy white to pearly white in color. The shell is thin and acutely pointed, with a flattened base. There are five or six whorls, encircled by strong revolving ribs, the upper ones often broken into a series of disjointed dots. The outer lip is thin, and is made wavy by the terminating ribs. There is no umbilicus. This is really a beautiful little shell, with a lustrous appearance both within and without, and very popular with shell collectors. It closely resembles the next species, *Margarites cinerea*, but it may be distinguished by its larger size, its luster, and by its complete lack of an umbilicus. The pearly top shell is found on gravelly bottoms in moderately shallow water, from Nova Scotia to Massachusetts.

Genus *Margarites* Leach 1847

18 Species and varieties

MARGARITES CINEREA Couthouy (Ridged Top Shell)

p. 129

This shell is nearly one-half inch high, and ashy white in color, sometimes tinged with green. The shell is thin and of a low pyramidal shape, with five to seven whorls that are rendered angular by prominent revolving ribs. The surface also bears very fine lines of growth. The umbilicus is broad and deep, and the aperture is circular, with a thin and sharp outer lip. This species also occurs on pebbly or gravelly bottoms, in shallow water just below the low water line. It is a cold water snail, ranging south to Cape Cod.

MARGARITES GRŒNLANDICA Gmelin (Wavy Top Shell)

p. 129

The wavy top shell averages about one-third of an inch in height and about one-half inch in diameter. Dull reddish brown in color, the shell is thin, with four or five whorls that are somewhat flattened above, and undulated near the sutures by short folds or wrinkles. The surface is sculptured by numerous elevated, revolving ribs. The umbilicus is broad and funnel-shaped, and the operculum is thin and horny. The inside of the aperture is pearly. This snail used to be listed as *Margarites undata*. It is a rather common species along the upper Maine coast. It may be seen occasionally on the rocks at low tide, but usually one needs to dredge in shallow water for specimens. The species provides an important food item for fishes from Cape Cod north.

MARGARITES HELICINA Phipps (Smooth Top Shell)

p. 129

This is a small shell, less than one-half inch across. Its color is pale yellowish green, often with an iridescent sheen. The shell is thin and translucent, smooth and shining. There are four or five whorls, the body whorl relatively large. The sutures are well impressed, and the surface is minutely scored by extremely fine growth lines. The aperture is circular, the lip thin and sharp, and the umbilicus is broad and deep. This little snail is found on eel-grass and other marine vegetation, well below the low-water level. After storms they may be found clinging to plants that have been torn from their moorings and cast up on the beach. This species ranges from Greenland to Massachusetts Bay.

MARGARITES OLIVACEA Brown p. 129
This is a diminutive snail, only about one-fourth inch in diameter. The olive-brown shell is thin and fragile, consisting of about four convex whorls, with distinct sutures. There is a deep umbilicus, and the inside of the aperture is quite pearly. This little fellow lives from Labrador to Cape Cod, and is generally a very common mollusk in the stomachs of fishes taken along the northern New England coast.

Genus *Solariella* Wood 1842

29 Species and varieties

SOLARIELLA INFUNDIBULUM Watson p. 201
This is a handsome little shell about three-fourths of an inch high, belonging to a group that contains a great many small gastropods. It is stoutly top-shaped, with about five rounded whorls tapering to a very sharp apex. The aperture is large and nearly round, the outer lip is thin and sharp, and there is a deep umbilicus. The shell is very thin and delicate, and each volution is decorated with sharp revolving lines on the lower part and strongly beaded lines on the upper part. The color is creamy white with a pearly iridescence. This species ranges from Newfoundland to North Carolina, living in deep water.

Family Phasianellidæ

CALLED "pheasant shells." The shell is fairly high-spired, and graceful in shape, porcellaneous, and polished. There is no periostracum. Tropical varieties attain a height of two inches and are among our most showy shells, usually brightly colored with pinks and browns. A very few members of this family occur on our shores, but they are all small.

Genus *Phasianella* Lamarck 1804

3 Species

PHASIANELLA AFFINIS Adams (Checkered Pheasant Shell) p. 129
The checkered pheasant shell is little more than one-fourth inch high. Its color is pale buff or gray, variously mottled and marked with pink and reddish brown, often with a narrow encircling band of yellow or orange on the body whorl. There are four or five volutions, the spire tapering gradually to a fairly sharp apex. The sutures are distinct. This colorful little snail may be found on seaweeds in southern Florida.

Family Turbinidæ

SHELLS GENERALLY heavy and solid, turbinate or top-shaped, and usually brightly colored with pearly underlayers. The surface may be smooth, rugose, or spiny. There is a heavy, calcareous operculum, and no umbilicus. These are herbivorous gastropods, native to tropic seas throughout the world. Many are used for ornamental purposes. The famous "cat's eye" of South Pacific Island jewelry is the colorful operculum of a member of this family (*Turbo petholatus* Linne).

Genus *Turbo* Linne 1758

4 Species and varieties

TURBO CASTANEUS Gmelin　　　　(Knobby Top Shell) p. 129
Sometimes known as the "chestnut top shell," this species is about one and one-half inches high, and buffy brown in color, more or less blotched with darker brown. Occasional specimens may be dull greenish. The shell is solid and roundly top-shaped, the apex sharply pointed. There are five or six whorls that are decorated with revolving bands of beads, those on the shoulders most pronounced. The aperture is large and round, and reflects somewhat on the columella. The operculum is thick and calcareous. The knobby top shell may be found from North Carolina to Mexico. It is a shallow water species, and should be looked for in grassy situations close to shore. The illustration on page 129 shows one shell with the operculum in place, and just above it a reversed operculum, showing the spiral growth that marks its brown inner surface.

TURBO CANALICULATUS Herman　　　　　　　　p. 204
This is a fine large gastropod, attaining a height of about two inches. There are some five well-rounded whorls, each one sculptured on its upper half by deeply incised revolving lines. The aperture is round, the outer lip sharp, and there is no umbilicus. There is a thick, round, calcareous operculum. The color is greenish yellow, usually mottled and checked with green and brown, and the shell's interior is pearly. This is an uncommon species in Florida, ranging south to Jamaica. It has been known for years as *Turbo spenglerianus* Gmelin.

Genus *Astræa* Bolten 1789

9 Species and varieties

ASTRÆA LONGISPINA Lamarck　　　　(Star Shell) pp. 29, 129
The star shell is a silvery white, iridescent shell, about two inches in diameter. The shell is strong and solid, with the

spire very low. The base is almost flat. There are six or seven whorls, and the margins are sharply keeled. The ornamentation consists of a series of triangular, sawlike spines which project beyond the periphery of the shell.

The star shell is by far the handsomest member of its group. Its silvery color and ornate appearance make it a great favorite with shell collectors and with the manufacturers of shell jewelry. It may be found on grassy bottoms in shallow water in southern Florida. Empty shells left exposed on the beach soon lose their luster. A very closely related species, *Astræa brevispina* Pilsbry, has shorter and less conspicuous spines, and is less abundant.

ASTRÆA AMERICANA Gmelin   (American Star)   p. 129
This shell, sometimes called the turbine shell, averages one inch in height, and is grayish white to greenish in color. The shell is solid and stony, with a well elevated spire. There are seven or eight whorls, with the apex bluntly rounded. The whorls are decorated with longitudinal folds which terminate in knobs on the shoulders. The outer lip is crenulate, and the operculum is calcareous. This is another herbivorous gastropod, found on seaweeds and algæ-covered rocks in shallow water. It is fairly abundant in the West Indies, and may be found as far north as central Florida. It occurs rather commonly as a fossil (with a white, chalky color) in the Pleistocene beds of Caloosahatchee, Florida.

ASTRÆA CÆLATA Gmelin   (Carved Star)   p. 29
This is a fair sized snail, attaining a length of three inches. Its color is greenish white, mottled with reds and browns. The shell is rugged and conical, and sculptured with a series of strong, oblique ribs which become tubular at the periphery, and equally robust revolving ridges. The aperture is large and oblique, and the columella is somewhat curved. There is a heavy calcareous operculum, dull white on the outside. This is a striking shell, much larger than the majority of its genus. It occurs in the West Indies and is taken occasionally at the Florida Keys.

ASTRÆA TUBER Linne   (Green Star)   p. 204
Some two inches in length, this is also a rugged and solid shell, consisting of about five whorls. It is sculptured with strong obliquely vertical ribs that are swollen at the base and at the shoulder to present a double row of nodes on each volution. The color is green, with the pearly layer generally showing through in many places, especially at the apex. The aperture is very pearly, and there is a broad, pearly area on the inner lip. There is no umbilicus, and the operculum is calcareous. This fine species is found on the reefs in Florida and the West Indies.

ASTRÆA IMBRICATA Gmelin   (Tubed Turbine)   p. 129
This shell is about one and one-half inches in height and greenish brown in color. The shell is strong and rough, and has about seven whorls, with very indistinct sutures. There are some twenty longitudinal folds on the body whorl, which form squarish, tubelike ribs, so that the surface is unusually rough and nodular. This is not a pretty shell. It is dull in color, and generally overgrown with bryozoans and sponges, and needs to be scraped clean in order to be at all attractive. It lives among the weeds and vegetation in moderately shallow water, from northern Florida south.

Family Neritidæ

SMALL, brightly colored snails, mostly globular in shape, and commonly with toothed apertures. They inhabit warm countries, where they are often found abundantly in shallow seas, in brackish waters, in fresh water, and even in some cases on dry land.

Genus *Nerita* Linne 1758

4 Species and varieties

NERITA PELORONTA Linne   (Bleeding Tooth)   p. 29
The bleeding tooth is one to one and one-half inches in diameter. Its color is yellowish white, marked with zigzag bars of red and black, with the black predominating. There is an orange-red stain on the columellar margin. The shell is thick and heavy, and semiglobose, with a very short spire. The surface bears several broadly rounded, revolving ribs. The outer lip is feebly toothed within, while the inner lip (columellar margin) sports one or two strong central teeth which are glistening white, surrounded by the rich orange stain. The operculum is shelly.
  Popular names are often misleading, but one has only to look into the aperture to see how perfectly the name "bleeding tooth" fits this gastropod. It is most active at night, when it roams about the rocks between the tide lines, feeding upon algæ. Extremely popular with shell collectors and visitors at our southern shores, the bleeding tooth may be found from Florida to the West Indies.

NERITA VERSICOLOR Gmelin   (Variegated Nerite)
p. 29
The variegated nerite is about one inch in diameter, frequently somewhat less, and is white in color, with zigzag bands of black and red. The interior is white. The shell is thick, porcellaneous, and semiglobose, with little if any spire. There are about four whorls, decorated with broad and round revolving ribs, with

narrow grooves between them. The outer lip is toothed within, and there are several robust teeth on the inner lip.

The gay colors of this species serve to distinguish from the next one, *tessellata*, while the pure-white inner lip separates it from the last, *peloronta*, so there is little difficulty in identifying this shell. It occurs from Florida to the West Indian Islands, and may be found rather commonly on the rocks at low tide.

NERITA TESSELLATA Gmelin (Checkered Nerite)
p. 29

This nerite averages between one-half and three-fourths of an inch in diameter, and is checkered black and white in color. Its shape is much the same as the two nerites just described, with four or five whorls separated by indistinct sutures. The surface is sculptured with about a dozen rounded spiral ribs on each volution, and with deep, narrow grooves between them. Numerous small teeth adorn the columellar margin, and, like the others, it has a shelly operculum.

This snail differs from the last one, *versicolor*, in being entirely black — no reds — and in usually being smaller. It occurs in rocky situations, often concealing itself in some crevice during the ebb tide, and is quite abundant on the west coast of Florida.

Genus *Purperita* Gray 1857

1 Species

PURPERITA PUPA Linne (Zebra Shell) p. 29

This striking shell is but one-half inch in diameter, often smaller. Its color is creamy white, spirally striped with fine, irregular black lines. The interior of the aperture varies from yellow to bright orange. The shell is thin, globular, with practically no spire, and it consists of only two or three whorls. The sutures are fairly distinct. The outer lip is thin and sharp, and there is a broad, flat, polished area at the base of the columella. The operculum is shelly, with a flexible border. This very attractive little snail is found in splash pools just above the high-tide mark, on the west coast of Florida, and in the West Indies. Large quantities of shells of this diminutive snail found concentrated in old Indian "kitchen middens" in Cuba suggest that the aborigines may have made a broth of them.

Genus *Neritina* Lamarck 1809

4 Species and varieties

NERITINA VIRGINEA Linne (Virginia Nerite) p. 29

The Virginia nerite is usually less than one-half inch in diameter. Its color is extremely variable, usually some shade of gray-green, tan, or yellow, scrawled all over with lines, circles, and dots of black. The shell is globular, consisting of three or four whorls, the large body whorl making up most of the shell. The aperture is oval, the outer lip thin, and the whole shell is highly polished. This is mainly a brackish water snail, but it occurs part way up rivers and streams, and dead shells are frequently washed up on marine beaches. The color patterns exhibited by this attractive species are almost endless, and a collection of a dozen specimens will show a startling and colorful variety. This species occurs from Florida to Brazil, and very similar species are found in tropic lands all around the world.

NERITINA RECLIVATA Say   (Green Nerite)   p. 29
This shell averages about one-half inch across, and is dark green in color, usually with fine black lines. The shell is globular, with three or four whorls, the body whorl constituting most of the shell. The aperture is large and oval, and the columellar area is pure white. The surface is smooth. This gastropod is fairly abundant on the Florida west coast. It inhabits brackish water, and is to be looked for in the tidal areas of streams. Many specimens lack the spiral black lines, and are a pure deep green, with a gleaming white inner lip.

Genus *Smaragdia* Issel 1869

1 Species

SMARAGDIA VIRIDIS Linne   (Green Neritina)   p. 204
Formerly classed as a *Neritina*, this is a very small shell, only about one-fourth of an inch across. The shape is rather globular and there are but two or three whorls, the last one making up most of the shell. The aperture is large, the outer lip sharp and thin, and there is a broad polished area at the base of the inner lip. The color is pale green, often with a few whitish streaks on the shoulder, and the surface is very shiny. This is a very abundant little gastropod in the West Indies, and great numbers can easily be gathered in the drift along shore in favorable localities. It is also found in southern Florida.

Family Epitoniidæ

THESE ARE predatory, carnivorous snails, occurring in all seas, and popularly known as "wentletraps" or "staircase shells." The shells are high spired, usually white and polished, and consist of many rounded, ribbed whorls which gradually decrease in size from the base to the summit. The outer lip is thickened considerably by a reflected border, secreted during rest periods in shell growth, and this thickened lip becomes a new riblike varix as the mollusk increases in size. The wentletraps are among the most delicately graceful of all marine mollusks.

Genus *Epitonium* Bolten 1798

58 Species and varieties

EPITONIUM PRETIOSUM Lamarck   (Precious Wentletrap)

p. 136

Probably the best known member of this group is the precious wentletrap. This is a stout, pure white shell some two inches high, occurring in deep water off the Asiatic coasts. This shell was once in great demand by collectors, and a dealer with good specimens could get almost any price he wanted to ask for them. The precious wentletrap, then known as *Scalaria pretiosa*, became so valuable that the clever Chinese made imitation shells out of hardened rice paste and flooded the European market with them. The shell is still far from common, but good specimens may be obtained from most dealers in conchological material for around five dollars each.

EPITONIUM LINEATUM Say   (Lined Wentletrap)

p. 136

The lined wentletrap is about one-half inch high, and pinkish white in color, sometimes with one or two brownish bands on the body whorl. The shell is elongate and high spired. There are about eight well rounded whorls, with well impressed sutures. Each whorl bears about sixteen delicate, slightly raised vertical ribs which do not cross the suture. The ribs are usually lacking on the lower part of the body whorl. The spaces between the ribs are smooth. This is a carnivorous snail, spending its time creeping over the ocean floor from well beyond the low-water line to abysmal depths. It is found commonly in the stomachs of cod, and, rarely, on the beach. It is a delicate and handsome shell, occurring from Massachusetts to Florida, most abundantly in the south.

EPITONIUM MULTISTRIATUM Say   (Crowded
Wentletrap)   p. 136

The crowded wentletrap is white in color and about one-half inch in height. The shell is highly elevated and acutely pointed, and has about eight rounded whorls, decorated with numerous moderately raised, equally spaced ribs. The aperture is round, with both the inner and outer lips thick and rounded. The operculum is horny. This glassy little shell has its ribs crowded somewhat closer together than with the others of its genus. It is found in rather deep water, all along the Atlantic coast. Occasional specimens are to be found in the drift alongshore, and many are taken from the stomachs of fishes.

EPITONIUM GRŒNLANDICUM Perry   (Ladder Shell)
                                                                p. 136
This wentletrap is a full inch in height, and white or yellowish
brown in color. The shell is elongate, tapering regularly to a
fine point. There are eight or nine somewhat flattened whorls,
barred with about a dozen stout, flattened, oblique ribs. The
spaces between the ribs are marked with rounded ridges and
revolving lines that follow the volutions of the shell. The
aperture is nearly circular, and is bordered by a stoutly thickened
lip which will, in turn, become another rib. This is one of
the showy shells of the north Atlantic coast, making up in
bizarre sculpture what it lacks in size. The decorative ribs
represent periods of rest from shell growth, when the mollusk
greatly thickened the rim of its aperture. Essentially a cold-
water species, the ladder shell is seen occasionally on the Maine
coast, and found quite commonly in the stomachs of fishes
taken off the Grand Banks.

EPITONIUM CENTIQUADRUM Morch   (Little Staircase
Shell)   p. 136
This little snail is nearly one-half inch high and glossy white in
color. The shell is thin and delicate, elongate, and has about
eight whorls, separated by distinct sutures. There are about
eleven vertical ribs on each volution, with the spaces between
smooth and polished. This little snail lives in rather deep water,
and most of the specimens obtained by collectors are recovered
from the stomachs of fishes. It occurs off the coasts of the
Carolinas.

EPITONIUM ANGULATUM Say   (Angled Wentletrap)
                                                                p. 136
The angled wentletrap is slightly less than one inch in height,
and is usually white in color. The shell is stoutly elongate,
with from six to ten whorls which do not touch each other in the
coil. There are about ten vertical, bladelike ribs on each
volution, each with a more or less blunt angle, or shoulder,
above, near the suture. It lives in deep water, and occurs from
Long Island to Florida, being most abundant in the south.

EPITONIUM EBURNEUM Potiez & Michaud   (Ivory
Wentletrap)   pp. 104, 136
The ivory wentletrap is fully an inch in height, and its color
is yellowish or ivory-white, usually with violet markings. The
shell is moderately thin but strong, high-spired, and sharply
pointed, and consists of about nine rounded whorls, with about a
dozen thickened ribs on each. The aperture is circular, and is
bordered by a thickened lip. This is a Florida species, usually
exhibiting more color than most of its group. It prefers to
live in rather deep water, but occasional specimens are to be
found on beaches after storms.

EPITONIUM HUMPHREYSII Kiener   (Humphrey's Wentletrap)   p. 136

This shell is a little more than one-half inch high, and white in color. It used to be listed as *Scalaria sayana* Dall. There are eight or nine rounded whorls, decorated with the usual vertical ribs. The spaces between the ribs are smooth and glossy, sometimes with faint spiral lines. This little shell is also an inhabitant of deep water, and occurs from Massachusetts to Texas.

EPITONIUM CLATHRUM Sowerby   (Trellis Wentletrap)
                                                         p. 136

The trellis wentletrap is about one inch tall, and white or creamy white in color. The shell is elongate, rather stout at the base, and thin and brittle. There are six or seven whorls, with eleven thin, bladelike ribs on each, and one revolving line at the base of the body whorl. The aperture is round, and the operculum is horny. This is a somewhat stouter shell than most of its group. It, too, is a deep water form, not very often seen on the beach. Its range is from Florida south.

EPITONIUM KREBSII Morch   (Kreb's Wentletrap)
                                                         p. 136

This is a little fellow, scarcely more than one-fourth inch in height. Its color is white. There are five or six convex whorls that are hardly joined at the sutures. The body whorl is relatively large, giving the shell a somewhat squat appearance. There are about ten ribs to a volution, slightly toothed at their upper ends. Kreb's wentletrap occurs in Florida and the West Indies.

Family Janthinidæ

THESE ARE floating mollusks (pelagic) living miles from land. Very delicate, lavender or purple shells, shaped much like land snails. There is no operculum. The animal is capable of ejecting a purplish fluid, which it does when disturbed.

Genus *Janthina* Bolten 1798

3 Species

JANTHINA JANTHINA Linne   (Violet Snail) pp. 104, 136

The violet snail is about one and one-half inches across. It is a two-toned shell, pale violet above and deep purple below. The graceful shell is thin and fragile, and consists of three or four sloping whorls. The surface bears very fine striæ. The outer lip is thin and sharp, and there is no umbilicus and no operculum.

THE TELLINS AND THE BORING CLAMS

1. *Tellina lintea* × 1, 2 views: **LINEN TELLIN**   p. 75
   Strong concentric ridges.

2. *Gastrochœna rostrata* × 1, 1 view   p. 94
   Small; squarish; yellowish white.

3. *Tellina magna* × 1, 2 views: **BIG TELLIN**   p. 78
   Large; often orange or red.

4. *Tellidora cristata* × 1, 2 views: **SAW–TOOTH**   p. 79
   Sharp projections on upper margin.

5. *Saxicava arctica* × 1, 2 views: **ARCTIC ROCK–BORER**
                                                                  p. 91
   Dull; squarish; surface rough.

6. *Zirfœa crispata* × 1, 1 view: **GREAT PIDDOCK**   p. 98
   White; gaping; prickly surface.

7. *Petricola pholadiformis* × 1, 2 views: **FALSE ANGEL WING**
                                                                  p. 72
   Elongate; thin-shelled; radiating ribs.

8. *Barnea truncata* × 1, 1 view: **TRUNCATED BORER**   p. 98
   White; gaping; thin-shelled.

9. *Pholas campechiensis* × 1, 1 view: **WING SHELL**   p. 95
   Very elongate; strong radiating ribs.

10. *Barnea costata* × ⅔, 1 view: **ANGEL'S WING**   p. 95
    White; thin-shelled but strong.

Plate 24   125

THE WOOD BORERS

1. *Panope bitruncata* × 1, 1 view   p. 91
   Large; white; smooth; gaping.

2. *Psammosolen cumingianus* × 1, 1 view   p. 86
   See text.

3. *Teredo navalis* × 1, 1 view: **COMMON SHIP WORM** p. 99
   Tubular galleries in wood.

4. *Martesia striata* × 1, 3 views: **WOOD PIDDOCK**   p. 99
   Thin-shelled; dark grayish.

5. Borings of *Martesia* × ⅔

This is the largest member of its genus to be found on our coast. It is a pelagic mollusk, living out its life many miles at sea, fastened to a float of its own making to which the eggs are also attached. It discharges a purple fluid when irritated. These snails float about in tremendous numbers, and sometimes huge rafts of them are blown ashore, where the purple color may stain the beach for considerable distances. Occasional empty shells are to be found on the Atlantic coast at almost any time, most commonly in the south.

JANTHINA GLOBOSA Swainson   (Globe Violet Snail)

p. 136

This is a smaller species, about three-fourths of an inch in diameter. Its color is pale violet all over, usually darker at the base. The shell is globular, thin, and very fragile, with about three rounded whorls, not sloping as in the last species. The aperture is large and moderately elongate. The globe violet snail is frequently found on Florida shores. Like its larger relative just discussed, it is a floating snail, at home in the warm currents of the Gulf Stream. Although the color is apt to be paler near the apex, it lacks the sharp dividing line between the whitish violet and the deep purple that is characteristic of *J. janthina*.

Family Melanellidæ

SMALL, HIGH-SPIRED shells, usually polished. The spire is often slightly bent at one side. This is a large family, living mostly in tropical waters. Many of the species are parasitic on other forms of marine life.

Genus *Melanella* Bowdich 1822

31 Species and varieties

MELANELLA INTERMEDIA Contraine   p. 137

This is an elongate shell about one-half inch high. Glossy white in color, there are from ten to thirteen whorls, tightly coiled and tapering very gradually to a sharp apex, with scarcely any perceptible sutures. The aperture is narrow, the outer lip thin and sharp, and the operculum is horny. This little gastropod occurs from New Jersey to the West Indies.

MELANELLA CONOIDEA Kurtz & Stimpson   p. 204

About three-fourths of an inch tall, this is a high-spired, shiny little fellow, composed of about twelve flat whorls. The sutures are quite indistinct, and the shell tapers very regularly to a pointed apex. The aperture is relatively small. The color is pure white, and the surface is highly polished. The range of this species is from Cape Hatteras to Florida.

Genus *Stilifer* Broderip 1832

5 Species

STILIFER STIMPSONI Verrill   (Stimpson's Stilifer) p. 201
This is an odd little snail, only about one-eighth of an inch high. There are four or five whorls, the last one forming a large part of the shell. The apex, or initial chamber, is relatively tall and slender, standing up like a miniature steeple. The surface of the shell is smooth, and the color is pale orange yellow. This little fellow is found most commonly as a parasite on the sea urchin, living among the many spines of the echinoderm. It may be collected from New Jersey north.

Genus *Aclis* Loven 1846.

18 Species and varieties

ACLIS WALLERI Jeffreys   (Waller's Aclis)   p. 137
This shell is about one-half inch tall, and glossy white in color. It has about ten whorls, fairly well rounded and with clearly defined sutures. The aperture is oval. This little shell occurs in deep water, from Labrador to Massachusetts.

Genus *Niso* Risso 1826

5 Species and varieties

NISO INTERRUPTA Sowerby   p. 137
This shell is about three-fourths of an inch in height, and its color is pale brown, sometimes with patches of darker brown. The sutures are marked with a reddish-brown line, and the whole surface is highly polished. The shell is conical, rather wide at the base, and consists of ten or eleven flattened whorls, so that the taper is very regular. This mollusk is said to be very active. It may be collected from Cape Hatteras to the Gulf of Mexico.

Family Pyramidellidæ

SMALL, pyramidal, or conical shells, usually white and polished. Many-whorled. The columella is usually plicate. They inhabit sandy bottoms, and the family contains a vast number of very small gastropods.

THE LIMPETS

1. *Acmæa testudinalis* × 1, 3 views: **TORTOISE-SHELL LIMPET**   p. 103
   No opening at summit; smooth.

2. *Lepeta cæca* × 4, 2 views   p. 106
   Tiny; well arched.

3. *Fissurella rosea* × 1, 2 views: **ROSY FISSURELLA**   p. 107
   Egg-shaped; commonly pinkish.

4. *Diodora listeri* × 1, 3 views: **KEY-HOLE LIMPET**   p. 110
   Opening at summit; strong network sculpture.

5. *Acmæa candeana* × 1, 2 views: **SOUTHERN LIMPET** p. 103
   No opening at summit; radiating ridges.

6. *Fissurella barbadœnsis* × 1, 2 views: **BARBADO LIMPET**
                                                             p. 106
   Opening at summit; usually green inside.

7. *Diodora cayenensis* × 1, 3 views: **LITTLE KEY-HOLE LIMPET**
                                                             p. 111
   White; hole at summit; strongly ridged.

8. *Fissurella nodosa* × 1, 3 views: **KNOBBY LIMPET**   p. 107
   Knobby; often rayed or banded.

9. *Lucapina sowerbii* × 1, 1 view: **CANCELLATED LIMPET**
                                                             p. 110
   Elongate opening at summit, usually stained at edge.

Plate 26   129

THE PEARLY SNAILS

1. *Tegula fasciata* × 1, 2 views   p. 111
   Smooth; rounded whorls; often colored.

2. *Calliostoma jujubinum* × 1, 2 views: **MOTTLED   TOP
   SHELL**   p. 112
   Top-shaped; no shoulders.

3. *Margarites grœnlandica* × 2, 2 views: **WAVY TOP SHELL**
                                                         p. 114
   Umbilicus; wavy surface.

4. *Calliostoma euglyptum* × 1, 2 views: **FLORIDA TOP SHELL**
                                                         p. 113
   Rounded shoulders; no umbilicus.

5. *Margarites cinerea* × 2, 2 views: **RIDGED TOP SHELL**
                                                         p. 114
   See text.

6. *Calliostoma occidentalis* × 2, 2 views: **PEARLY TOP SHELL**
                                                         p. 113
   No umbilicus; often pearly.

7. *Margarites olivacea* × 2, 2 views   p. 115
   Very small; grayish.

8. *Margarites helicina* × 2, 2 views: **SMOOTH TOP SHELL**
                                                         p. 114
   See text.

9. *Turbo castaneus* × 1, 3 views: **KNOBBY TOP SHELL**
                                                         p. 116
   Strong and solid; beaded shoulders.

10. *Phasianella affinis* × 2, 2 views: **CHECKERED PHEASANT
    SHELL**   p. 115
    Small; pinkish, often banded.

11. *Astræa imbricata* × 1, 1 view: **TUBED TURBINE**   p. 118
    Stoutly conical rough.

12. *Astræa americana* × 1, 2 views: **AMERICAN STAR**   p. 117
    Conical; fluted whorls.

13. *Astræa longispina* × 1, 2 views: **STAR SHELL**   p. 116
    Pearly; flattish.

Genus *Pyramidella* Lamarck 1799

21 Species and varieties

PYRAMIDELLA DOLABRATA Linne (Obelisk Shell)

p. 204

This is an attractive shell, almost one inch high. There are nine or ten rounded whorls, with sutures that are deeply impressed. The taper is rather abrupt, and the small shell is quite plump. The aperture is semi-lunar, the outer lip thin and sharp, and there are two or three distinct ridges on the inner lip. There is a small umbilicus. The color of the shell is pale yellow, and the surface is highly polished. This species is found in southern Florida.

PYRAMIDELLA CRENULATA Holmes p. 137

This species is about one-half inch in height, and is pale brown in color, often with the columella darker brown. There are about a dozen rather flat whorls, with channeled sutures. The operculum is corneous, brown in color, semicircular in outline, and is notched on one side to fit the plaits on the columella. This is a delicate little shell, shiny and somewhat translucent. It may be found on sandy bottoms in shallow water, from South Carolina to the Gulf of Mexico.

Genus *Turbonilla* Risso 1826

74 Species and varieties

TURBONILLA STRIATULA Couthouy p. 137

The genus *Turbonilla* is a large one, containing many species of very small, high-spired gastropods, all very much alike to the casual observer. This is one of the commoner forms, about one-half inch tall, that may be found from Nova Scotia to Massachusetts. There are some nine or ten whorls, with well-impressed sutures. The shell is sculptured with vertical folds, which are nearly lacking on the specimen shown on page 137.

TURBONILLA HEMPHILLI Bush (Hemphill's Turbonilla)

p. 204

This is a tall and slender, high-spired snail with a height of nearly one-half inch. There are about fifteen whorls, with distinct sutures, and each whorl is decorated with a number of conspicuous vertical ribs. The aperture is moderately small. The color is white. This rather pretty little shell occurs on the Florida west coast.

TURBONILLA RATHBUNI Verrill & Smith (Rathbun's Turbonilla

p. 204

This is one of the largest members of its genus, attaining a length of well over one-half inch. There are about twelve whorls, with deep sutures, and the color is white or buffy white. The vertical ribs, so characteristic of this group, are small and rather crowded. This species is found from Martha's Vineyard to North Carolina.

TURBONILLA ELEGANTULA Verrill   (Elegant Turbonilla)
p. 204
This is a tiny shell, not much more than one-eighth of an inch long. The shape is slender, with about ten whorls, each sculptured with conspicuous vertical folds, or ribs. The sutures are distinct, and the aperture is small and oval. The color is yellowish white, and the surface is shiny. This diminutive univalve lives in moderately deep water, off the southern New England coast.

Genus *Odostomia* Fleming 1817
30 Species and varieties

ODOSTOMIA TRIFIDA Totten   p. 137
This is a tiny, pale greenish-white shell, about one-fifth of an inch long. The shell is conical, and has five or six whorls, each one separated by a well defined suture. There are two or three impressed lines below the suture, giving the shell the appearance of having a multiple suture. The aperture is oval, the outer lip simple, and the inner lip bears a single, oblique fold. This little mollusk is found under stones and pieces of driftwood at low tide. Dead shells can usually be found by carefully searching the line of flotsam and jetsam that marks the high-tide limit on sandy beaches. This species occurs from Massachusetts to New Jersey.

ODOSTOMIA BISURTALIS Say   p. 137
This shell is about one-fourth of an inch high, and its color is dull white, with a thin, brownish periostracum. There are seven or eight whorls, somewhat flattened and slightly shouldered at the sutures. There is a deeply impressed revolving line on the upper part of each volution. The aperture is oval, and the columella bears one oblique fold. The operculum is brown and horny. This little fellow inhabits sandy mud flats, from the Gulf of St. Lawrence to Florida.

ODOSTOMIA IMPRESSA Say   p. 137
This is about the same size as the last species, one-quarter inch high, and it is milky white in color. The shell is elongate, consisting of six or seven rather flattened whorls, with channeled sutures. The surface is decorated with three equally spaced spiral grooves. The aperture is oval, with the outer lip thin and sharp, occasionally flaring a little in old shells. This species ranges from southern New England to the Gulf of Mexico, and is commonly found on oyster beds

ODOSTOMIA SEMINUDA Adams   p. 137
 This shell is also about one-fourth of an inch high, and its color
 is white, often glossy. It is stoutly conical, and has six or
 seven whorls, with distinct sutures. The shell is sculptured with
 several revolving ridges, and cut by vertical striations, so that
 the surface appears beaded in many cases. There is a strong
 oblique fold, or plait, at the base of the columella, and within
 the aperture. This gastropod occurs from Prince Edward Island
 to the Gulf of Mexico.

Family Naticidæ

COMMONLY KNOWN as "moon shells" or "shark eyes," these are
carnivorous snails, found in all seas. The shell is usually globular,
sometimes depressed, smooth, and polished. The foot of the
mollusk is very large, and often conceals the entire shell when the
animal is extended.

Genus *Natica* Scopoli 1777
10 Species and varieties

NATICA CANRENA Linne   (Spotted Moon Shell)

p. 29
 This shell is from one to two inches in height, and pale bluish
 white in color, with spiral chestnut bars, stripes, and zigzag
 markings, along with various mottlings of dark brown and
 purple. The shape is globular, somewhat flattened at the top,
 with a smooth and shining surface. The body whorl is large and
 expanded, the apex small, and the spire depressed. The umbilicus is deep and partly filled with a shiny white callus. The
 operculum is calcareous, white, thick, and it has several deeply
 cut grooves or channels following its outside curvature. The
 colorful spotted moon shell prefers sandy bottoms, where it
 may be found burrowing just beneath the surface of the sand at
 the low-tide mark. It feeds upon other mollusks, enveloping the
 smaller ones in its oversize foot and drilling into the shells
 of the larger varieties. It has been observed eating dead fish,
 but it is not ordinarily considered a scavenger. Shells of this
 snail are rather common objects on the beaches from North
 Carolina to the West Indies, but most of them are apt to be faded
 and worn, and one needs to see a living specimen in order to
 appreciate the delicate coloring of this gay moon shell.

NATICA MAROCCANA Dillwyn   (Cat's-eye)   p. 140
 The little cat's-eye averages about one and one-half inches in
 height. It is rich brown in color, with the aperture and columella
 pure white. The shell is strong and solid, with four or five

whorls, the body whorl very large. The white inner lip is thick and heavy and partially covers the umbilicus. The operculum is calcareous, with a double marginal rib. This little natica may be found on the sand and mud flats from North Carolina to Florida. Like the others of its group, it is a carnivorous snail, burrowing just beneath the surface of the mud in search of prey, chiefly bivalves. When a clam is encountered, the snail quickly envelops it in its huge foot and proceeds to bore a neat round hole through its shell.

NATICA CLAUSA Broderib & Sowerby (Northern Natica)
p. 140

This little moon shell gets to be an inch and a half high, but most of the shells found will average less than one inch. Its color is pale brown. The shell is subglobular, consisting of four or five whorls, the spire only slightly elevated. The aperture is oval, and the outer lip is sharp, thickened and rounded as it ascends to the umbilicus, which is completely closed by a shiny, ivory-white callus. The operculum is calcareous, and bluish white in color. This is a cold-water species, found commonly in the stomachs of cod and other fishes taken off the Newfoundland and Nova Scotia coasts, and occasional on the tidal flats of Maine.

Genus *Polinices* Montfort 1810
15 Species and varieties

POLINICES HEROS Say (Common Northern Moon Shell)
p. 140

This is a very common and well-known shell on our northern beaches, being abundant from New Jersey to Maine. It gets to be four inches in height, and is ashy brown in color. The shell is thick, globular, and has a thin, yellowish brown periostracum. There are about five very convex whorls, somewhat flattened at the top. The aperture is large and oval, and the operculum is horny. The umbilicus is large, rounded, coarsely wrinkled, and extends through to the top of the shell.

This, and the next species, were originally described as naticas, but since the genus *Natica* calls for a shell with a calcareous operculum, these forms with horny or corneous opercula do not fit. At various times they have been called *Lunatia* and *Neverita* as well as *Natica*, until eventually the genus *Polinices* was created for them.

Polinices heros burrows in moist sand, its whereabouts at low tide being indicated by a small mound projecting above the general surface of some sand bar. The foot is tremendous in size, and one wonders that the creature can withdraw entirely into its shell. It is a voracious feeder, drilling into, and sucking the contents from, many an unlucky bivalve encountered in its subterranean wanderings.

POLINICES DUPLICATA Say (Lobed Moon Shell)

p. 76

This is another common moon shell, about two inches in height but nearly three in diameter. Its color is chestnut-brown, tinged with bluish. The shell is solid and oval, the upper portion of the whorls compressed so as to give it a somewhat pyramidal outline. There are four or five whorls. The aperture is oval and oblique, and the operculum is horny. The umbilicus is irregular, and is covered wholly or in part by a very thick, chestnut-brown callus. This species has a much flatter appearance than *heros*, and is somewhat smaller, but it may be instantly recognized by the brown lobe of shelly material extending over the umbilicus.

The eggs of this species, as with the last one, are laid in a mass of agglutinated sand that is molded over the shell, and, upon hardening, form the fragile "sand collars" often found lying on the beach in the summer months. This moon shell lives in company with the larger *heros*, upon the mud and sand flats at low tide, but its range extends much further south, it being a common shell from Maine to Florida.

POLINICES TRISERIATA Say

p. 140

This is a colorful little moon shell, less than one inch in height, and generally very much less. Its color is yellowish gray, sometimes uniformly so, and sometimes with oblique chestnut-brown squarish spots arranged in bands. There are three revolving bands on spots on the body whorl, and usually one on each of the upper whorls, although the whole shell may be unspotted. There are four or five whorls, and the general shape is almost exactly like a miniature specimen of *P. heros*; in fact, some writers have contended that *triseriata* is indeed merely the juvenile stage of *heros*, but certain characters, particularly a relatively thick callus on the inner lip, seem to indicate a mature shell. This gayly colored little fellow lives from Connecticut north.

POLINICES LACTEA Guilding (Milky Moon Shell)

p. 140

This is a milky-white gastropod about one inch in height. The shell is obliquely oval, smooth and polished, and less flattened than most of its genus. It bears a thin, yellowish periostracum during life. There are three or four whorls, the last one greatly enlarged. The umbilicus is partly filled by a callus, and the operculum is corneous, amber or claret in color. This shiny little shell may be found on the beaches from Florida to Texas. It normally lives out beyond the low tide limits, where it burrows an inch or so deep in the sand, but its empty shell, usually minus the periostracum, is commonly washed ashore.

Genus *Amauropsis* Morch 1857

1 Species

AMAUROPSIS ISLANDICA Gmelin   p. 201
About one inch in height, this is a rather plump shell of four whorls. Thin in substance, the shell is yellowish white in color, with a periostracum that is orange-brown. The aperture is semicircular, and about two-thirds the length of the shell. The outer lip is thin and sharp, with the inner lip overspread with a white callus. There is a small, slitlike umbilicus. This snail lives in rather deep water, from Massachusetts north.

Genus *Bulbus* Brown 1839

1 Species

BULBUS SMITHII Brown   p. 201
This little snail looks like a small, thin, *Polinices*. It is rather globular in shape, with a low apex, and is composed of about three whorls. Its length is usually under one inch, and its color is greenish yellow, with a thin periostracum, the interior dull white. There is no umbilicus. This is another deep-water gastropod, living in cold seas. It ranges from the Gulf of St. Lawrence to Maine.

Genus *Sinum* Bolten 1798

3 Species

SINUM PERSPECTIVUM Say   (Ear Shell)   p. 140
The ear shell is about one and one-half inches in diameter and less than one-half inch high. The milky-white shell is ovate-elongate, and greatly depressed. There are three or four whorls. The surface is sculptured with numerous impressed, transverse, slightly undulating lines. The body whorl is enormously expanded, the aperture being wide and flaring, making up more than three-fourths of the entire area of the shell. There is no umbilicus, and the operculum is rudimentary.

This curious gastropod looks for all the world like a *Polinices* that has been squeezed flat. It occurs on sand bars in shallow water, rather sparingly north of the Jersey beaches, but abundantly south of Cape Hatteras.

Family Lamellariidæ

SMALL, THIN-SHELLED mollusks, living chiefly in cold seas, and usually at considerable depths. Most of them feature a very heavy periostracum.

THE WENTLETRAPS

1. *Epitonium lineatum* × 2, 2 views: **LINED WENTLETRAP**
   p. 121
   Small; white and glossy.

2. *Epitonium multistriatum* × 2, 2 views: **CROWDED WEN-
   TLETRAP**
   p. 121
   Ribs closely crowded.

3. *Epitonium humphreysii* × 2, 2 views: **HUMPHREY'S WEN-
   TLETRAP**
   p. 123
   See text.

4. *Epitonium clathrum* × 2, 2 views: **TRELLIS WENTLETRAP**
   p. 123
   See text.

5. *Epitonium krebsii* × 2, 1 view: **KREB'S WENTLETRAP**
   p. 123
   Small and squat.

6. *Epitonium grœnlandicum* × 2, 2 views: **LADDER SHELL**
   p. 122
   Revolving grooves between ribs.

7. *Epitonium angulatum* × 2, 1 view: **ANGLED WENTLE-
   TRAP**
   p. 122
   Ribs form sharp angles at shoulders.

8. *Epitonium pretiosum* (**CHINA**) × 1, 2 views: **PRECIOUS
   WENTLETRAP**
   p. 121
   See text.

9. *Epitonium centiquadrum* × 2, 2 views: **LITTLE STAIRCASE
   SHELL**
   p. 122
   See text.

10. *Epitonium eburneum* × 3, 1 view: **IVORY WENTLETRAP**
    p. 122
    Commonly with lilac marks.

11. *Lacuna vincta* × 5, 2 views: **LITTLE CHINK SHELL**
    p. 150
    Elongate chink beside columella.

12. *Janthina globosa* × 1, 2 views: **GLOBE VIOLET SHELL**
    p. 126
    Globular; solid purple.

13. *Janthina janthina* × ½, 3 views: **VIOLET SNAIL**   p. 123
    Flatter than 12; pale violet above, darker below.

Plate 28   137

THE PYRAMID SHELLS AND SOME OTHERS

1. *Xenophora trochiformis* × ½, 3 views: **CARRIER SHELL**
   p. 138
   With attached shell fragments.

2. *Pyramidella crenulata* × 2, 2 views   p. 130
   Flat whorls; glassy white.

3. *Aclis walleri* × 3, 2 views: **WALLER'S ACLIS**   p. 127
   Elongate; glassy; whorls rounded.

4. *Odostomia trifida* × 4, 2 views   p. 131
   Impressed lines below sutures.

5. *Melanella intermedia* × 3, 2 views   p. 126
   Many-whorled; glassy.

6. *Odostomia bisurtalis* × 4, 2 views   p. 131
   Dull white; one impressed line below suture.

7. *Odostomia seminuda* × 4, 2 views   p. 132
   Cross-hatched sculpture.

8. *Odostomia impressa* × 4, 2 views   p. 131
   See text.

9. *Separatista cingulata* × 3, 2 views   p. 146
   Minute; flatly coiled.

10. *Turbonilla striatula* × 4, 1 view   p. 130
    With fluted whorls.

11. *Niso interrupta* × 4, 2 views   p. 127
    Tapering regularly; no shoulders.

Genus *Velutina* Fleming 1822

2 Species

VELUTINA LÆVIGATA Linne   p. 201
About one-half inch in length, this is a thin and fragile shell
of about three whorls. The color is pale brown or light tan.
The shell is nearly all body whorl, with a large and flaring aperture. The surface is rather smooth, and there is a thick, brownish periostracum that is usually ragged and frayed. This gastropod gives the impression that it is made up of the periostracum, with a very thin layer of shell on the inside. This
species lives in rather deep water, but it is occasionally found
on the beach. Its range is from Cape Cod north.

Family Xenophoridæ

SHELL top-shaped, considerably flattened. All but the deep-sea
forms camouflage their shells by cementing pebbles and broken
shell fragments to them, so that from above they look like small
piles of debris. Called "carrier shells," they are inhabitants of
warm seas, chiefly in the Pacific and Indian Oceans.

Genus *Xenophora* Waldheim 1807

3 Species

XENOPHORA TROCHIFORMIS Born   (Carrier Shell)
                                                               p. 137
This shell is about two inches in diameter, and yellowish brown
in color. The apex is low but sharp, and the volutions are
flattened. The surface bears prominent growth lines, although
they are usually visible only from the lower side. The aperture
is large and oblique, the outer lip thin. There is no umbilicus,
and the operculum is corneous.

This snail has a most extraordinary habit. Early in life it
fastens a bit of shell or a tiny pebble to its back, at the upper
edge of its aperture. As the snail grows the foreign object is
firmly anchored to its shell, and a larger piece is then added.
This results in a spiral of foreign material following the suture
line, the purpose of which is apparently concealment. Shells,
pebbles, or bits of coral may be used, but the mollusk usually
sticks to one kind of material, and rarely mixes them.

Similar carrier shells occur in the tropical seas of China and
Java. This species, occurring from North Carolina to Cuba,
seems to prefer bivalve shells, particularly those of *Chione cancellata*, and it is noteworthy that the bits of shell are so disposed
as not to curve downwards beyond the edge of the shell, which

would impede the progress of the animal, but are usually placed with their concave sides uppermost. Fossil specimens of this genus are found as far back as the Cretaceous period (one hundred million years), showing that this curious habit of self-ornamentation has persisted for countless thousands of generations.

Family Capulidæ

SHELL CONICAL, cap-shaped, without internal plate or cup. The apex is spiral, the base is open, and the whole shell is considerably curved.

Genus *Capulus* Montfort 1810

2 Species

CAPULUS UNGARICUS Linne   (Cap Shell)   p. 140
The cap shell is from one to one and one-half inches high and about one inch across at the base. The conical shell reminds one of a jester's hat, the apex coiled and curved forward. The gray surface bears fine, close, radiating lines, which are crossed by less frequent lines of growth. A periostracum, often hairy, covers the shell during life. The open base is shaped to fit the object to which the gastropod is attached, for the cap shell is limpet-like in its habits. It forms a shallow excavation at the place of attachment, and sometimes deposits a shelly floor. The cap shell is a modern representative of a very ancient group, fossil specimens of the same genus being found in rocks of the Silurian age. Our species lives in moderately deep water, commonly on oyster beds, all the way from Greenland to the West Indies.

Family Hipponicidæ

SHELLS THICK and obliquely conical, the apex hooked backward, but not spiral. The surface is generally rough and grayish white. The animal secretes a shelly plate between itself and the object on which it lives, and this plate was once believed to be a second valve, and several species were described as bivalve mollusks.

Genus *Hipponix* Defrance 1819

3 Species

HIPPONIX ANTIQUATA Linne   (Hoof Shell)   p. 140
The hoof shell is about one-half inch in height, and white or

THE MOON SHELLS

1. *Natica maroccana* × 1, 2 views: **CAT'S–EYE**   p. 132
   Rusty brown; large umbilicus.

2. Sand collar of *Polinices heros*   pp. 133, 134

3. *Polinices lactea* × 1, 2 views: **MILKY MOON SHELL**
   p. 134
   Milky white, glossy; small but solid.

4 *Polinices triseriata* × 1, 1 view   p. 134
   Small; usually with chestnut marks.

5. *Natica clausa* × 1, 2 views: **NORTHERN NATICA**   p. 133
   Rather thin-shelled; dull yellowish gray.

6. *Polinices heros* × 1, 1 view: **NORTHERN MOON SHELL**
   p. 133
   Large and globular; deep umbilicus.

7. *Sinum perspectivum* × 1, 3 views: **EAR SHELL**   p. 135
   White; very flat.

8. *Hipponix antiquata* × 2, 2 views: **HOOF SHELL**   p. 139
   White; rough and irregular.

9. *Capulus ungaricus* × 1, 2 views: **CAP SHELL**   p. 139
   Shaped like a jester's cap.

10. *Crucibulum striatum* × 1, 3 views: **CUP–AND–SAUCER LIM-
    PET**   p. 142
    With cuplike structure within.

Plate 30   141

THE SLIPPER SHELLS AND THE WORM SHELLS

1. *Crepidula convexa* × 1, 3 views: **LITTLE BOAT SHELL**
   p. 143
   Small; dark brown; deeply cupped.

2. *Crepidula aculeata* × 1, 2 views: **THORNY SLIPPER**   p. 144
   Rather flat; thorny.

3. *Crepidula fornicata* × 1, 3 views: **BOAT SHELL**   p. 143
   Strong and solid; irregular in shape.

4. *Crepidula plana* × 1, 3 views: **FLAT SLIPPER**   p. 144
   Milky white; very flat.

5. *Crepidula plana* on snail shell × ½

6. *Cheilea equestris* × 1, 1 view   p. 142
   Horseshoe-shaped plate inside.

7. *Cæcum pulchellum* × 20, 2 views   p. 153
   Minute; tubular; ringed.

8. *Serpulorbis decussatus* × 1, 1 view   p. 152
   Contorted tubes.

9. *Vermetus irregularis* × 1, 1 view   p. 152
   Tangled tubular masses.

grayish white in color. The shell is thick, conical, cap-shaped, and concave at the base. The apex is bluntly pointed and curved backward, but it is not coiled as in the cap shell just described. The surface is variable, sometimes fairly smooth, but often wrinkled by coarse laminations. There is a hairy periostracum. The little hoof shell is variable in shape, as it lives attached to some rock or dead shell, and grows to conform with its particular shape of seat, secreting a calcareous plate between itself and the object to which it adheres. It is found offshore in southern Florida.

Family Calyptræidæ

LIMPETLIKE gastropods, cup-shaped, with a cuplike process on the inner side of the shell, living attached to other shells. Found in all seas, from shallow water to moderate depths.

Genus *Cheilea* Modeer 1793

2 Species

CHEILEA EQUESTRIS Linne   p. 141
This shell is about one inch in diameter, and pale brown or gray in color. The shell is almost orbicular in outline at the base, and it rises to a blunt apex that is situated somewhat posterior in position and directed backwards. The exterior is closely rayed with distinct ridges, and the edge of the shell is lightly crenulated. Inside the shell is a sturdy, horseshoe-shaped plate. This species occurs from North Carolina to Texas and the West Indies. It lives attached to other shells, and usually excavates a shallow cavity at the place of attachment. *Calyptræa centralis* Conrad is a very similar shell, white in color, that occupies the same geographic range. It is a little fellow, only about one-quarter of an inch in diameter.

Genus *Crucibulum* Schumacher 1817

2 Species

CRUCIBULUM STRIATUM Say   (Cup-and-saucer Limpet)
                                                                    p. 140
This shell is about one inch in diameter, and pinkish white in color, generally streaked with brown. It is moderately solid and conical, the surface bearing numerous slightly elevated, radiating lines. The summit is usually smooth and bluntly pointed. The inner partition is cup-shaped and attached by one side to the shorter end of the shell. This small cuplike structure makes the popular name particularly appropriate for this gastropod. It

may be found clinging to rocks just below the low-water line, all along the Atlantic coast.

Family Crepidulidæ

SHELL generally oval, arched, and more or less boat-shaped. The inner cavity is partially divided by a horizontal platform. Adults are always fixed to some solid object, probably for life, although they may be able to move about slightly. The shape of the shell varies according to the object to which it is attached. World-wide in distribution, in warm and temperate seas.

Genus *Crepidula* Lamarck 1799
5 Species and varieties

CREPIDULA FORNICATA Linne   (Boat Shell)   p. 141
Known as the "boat shell" or "quarter-deck," this species is about one and one-half inches long, and its color is soiled white, flecked with purplish chestnut. The shell is obliquely oval, with the apex prominent and turned to one side, not separated from the body of the shell. It is moderately convex, according to the object on which it is seated. The cavity is partially divided within by a horizontal white platform called the diaphragm.

These are among the first objects collected by children at the seashore, as they make excellent miniature boats to sail in quiet tide pools, and also serve as tiny scoops for digging in the sand. The empty shells also have a commercial value, as, under the trade-name of "quarter-decks," many tons are annually scattered over the ocean floor for embryo oysters to settle upon. This is the largest and commonest *Crepidula* found on our coast. They are usually found alive adhering to each other and to other shells, particularly oysters, and their empty shells are strewn along the beaches from Nova Scotia to the Gulf of Mexico.

CREPIDULA CONVEXA Say   (Little Boat Shell)   p. 141
The little boat shell, or "slipper shell," is less than one-half inch long, and is ashy brown in color, with streaks and dots of dark, reddish brown. The shell is small, obliquely oval, and deeply convex. The apex is prominent and separate from the body of the shell, turning down nearly to the plane of the aperture, occasionally beyond it. The shelly partition (diaphragm) is deeply situated. The outer surface of the shell is minutely wrinkled. This species is not as abundant as the one just described, but it is still a very common species all along the

Atlantic coast. It is found on shells and attached to stones and seaweeds. It has been introduced in California waters along with oysters.

CREPIDULA ACULEATA Gmelin   (Thorny Slipper)

p. 141

The thorny slipper is about one inch in length, and brownish or purplish in color, quite variable in its markings. Commonly there are broad rays of a paler tint. The apex is turned considerably to one side. The outer surface bears irregular, thorny, or spiny ridges, radiating from the apex. The interior is polished and often rayed with brown. The diaphragm is white. This is a common snail along the shore from Cape Hatteras to the West Indies.

CREPIDULA PLANA Say   (Flat Slipper)   p. 141

The flat slipper shell is pure white, about one inch long, and is ovate, thin, and semitransparent, the outer surface wrinkled with concentric lines of growth. The inner surface is highly polished, sometimes iridescent. The apex is pointed and turned a little to one side. The platform is less than half the length of the shell. The general shape may be flat, slightly convex, or even concave, according to the outline of the object to which it is attached. The flat slipper is to be found flattened against the inside of the aperture of dead shells, particularly large specimens of *Busycon* and *Polinices*. Often a large king-crab, or horseshoe crab (*Limulus*), will be found to have several dozen on the lower side of its shell. This species inhabits the Atlantic coast from Maine to Florida.

Family Rissoidæ

VERY SMALL snails, living for the most part on stones, shells, and sponges. They are variable in sculpture, and occur in nearly all seas. Hundreds of species have been described.

Genus *Alvania* Risso 1826

21 Species

ALVANIA CARINATA Mighels & Adams   p. 188

This is a very tiny shell, about one-tenth of an inch high, and brown in color. The shell is moderately conical and very thin, consisting of about five whorls that are quite convex. The upper whorls are marked with longitudinal ribs, while the lower half of the body whorl bears revolving lines. The aperture is nearly circular. This is a cold-water gastropod, occurring from Maine north.

ALVANIA JAN-MAYENI Friele   p. 188
This is also a tiny brown snail, averaging about one-eighth of an inch in height. There are four or five whorls, with distinct sutures. There is a sculpture of prominent revolving ribs on each volution, the one on the shoulder broken into beads or knobs, giving the shell a decorative appearance, in spite of its diminutive size. This species may be found in moderately deep water, from the Gulf of St. Lawrence to North Carolina.

Genus *Rissoina* Orbigny 1840

8 Species

RISSOINA FENESTRIATA Schwartz   p. 188
This shell averages less than one-fourth inch in height and is white in color. The shell is solid, high-spired, and contains seven or eight whorls. There is a sculpture of strong longitudinal ribs crossed by equally prominent spiral ribs, so that the shell has a beaded appearance. The aperture is oval, and the outer lip crenulate. This small snail occurs in Florida and the West Indies.

RISSOINA CANCELLATA Philippi   p. 188
This is another small, rather elongate species, about one-fourth of an inch tall. There are about six whorls, well rounded and decorated with both longitudinal and revolving ribs, so that the surface is strongly cancellated, as the specific name implies. This gastropod lives in moderately shallow water, on seaweeds, sponges, etc., in southern Florida.

Genus *Hydrobia* Hartmann 1840

4 Species

HYDROBIA MINUTA Totten   p. 188
This is a small, shiny, yellowish-brown snail, about one-fourth of an inch in height. The shell is thin and semitransparent, and has about five rounded whorls, the surface faintly wrinkled by growth lines. The aperture is oval and about one-third the total height of the shell. There is a horny operculum. These diminutive snails are found plentifully on seaweeds and grassy banks, at about the high tide level, especially in ditches and brackish water pools about marshes. They are quite active, and move about with agility. This species occurs from New England to the New Jersey beaches.

Family Skeneidæ

VERY TINY, flattish snails, with few whorls, found under stones in moderately shallow water. Widely distributed.

Genus *Skenea* Fleming 1828

1 Species

SKENEA PLANORBIS Fabricius   p. 188
This is an extremely minute shell, about one-tenth of an inch in
diameter. It is dull brown in color, and rather flatly coiled, with
four whorls and very little spire. The sutures are well defined,
and the umbilicus is relatively deep and wide. These little
snails may be found living on oysters, sponges, corals, and all
sorts of objects in the sea, from Greenland to Florida.

Genus *Separatista* Gray 1847

1 Species

SEPARATISTA CINGULATA Verrill   p. 137
This is another rather flatly coiled little snail, about one-eighth
of an inch across. It is grayish white in color and has three or
four whorls. The aperture is round, the lip thin and sharp, and
there is a small umbilicus. There is a series of rather weak
revolving lines on each volution. This is a deep water mollusk,
found from Georges Bank to Delaware Bay.

Family Architectonicidæ

SHELL solid, circular, and but little elevated. The umbilicus is
very broad and deep, and is bordered by a knobby keel. Called
"sundial shells," they are confined to warm seas.

Genus *Architectonica* Bolten 1798

4 Species

ARCHITECTONICA GRANULATA Lamarck   (Sundial)
pp. 104, 173
The sundial shell is about one and one-half inches in diameter.
Its color is white or gray, spotted and marbled with brown and
purple. The shell is circular and somewhat flattened, with six
or seven whorls and rather indistinct sutures. The base is flat.
The surface is finely checked by crossing spiral lines and radiating ridges, producing a pattern of raised granules. The umbilicus is wide and deep, and strongly crenulate. The aperture
is round, and the outer lip is thin and sharp. This distinctive
gastropod may be found in moderately shallow water, especially
during the summer months, from North Carolina to the West
Indies. It was formerly known as *Solarium verrucosum* Philippi.

Related species, some of them much larger but otherwise almost identical, are found in the Pacific and Indian Oceans.

Family Littorinidæ

A LARGE FAMILY of shore-dwelling snails, found clinging to the rocks between the tide marks, and sometimes well beyond the average high-tide limits. The shell is usually sturdy, has few whorls, and is without an umbilicus. The distribution is world-wide. Commonly called periwinkles.

Genus *Littorina* Ferussac 1822

12 Species and varieties

LITTORINA LITTOREA Linne   (Common Periwinkle)

p. 156

The common periwinkle of northern rocky shores, this snail is from one-half to one inch in height, and brownish olive to nearly black in color, usually spirally banded with dark brown. The shell is heavy and solid, with six or seven whorls. The outer lip is thick, and black on the inside, and the base of the columella is white. There is no umbilicus. The apex is sharp, but the shell appears rather squat.

This species might be called the English sparrow of the mollusks. Probably the most abundant and best-known snail on the north Atlantic coast, this periwinkle is an immigrant from European waters, having been introduced (accidentally, probably in the egg stage) in the vicinity of Nova Scotia. The first specimen was found there in 1857, and from that date they have spread down the coasts of northern states as far as New Jersey, where their southern migration has been halted, probably by the temperature of the water, and also possibly by the sandy Jersey beaches. In Europe these dingy little snails are roasted and sold from pushcarts in the city streets. In America there seems to be an aversion to eating them, though just why we should consider clams, oysters, and scallops as delicacies, and at the same time be horrified at the thought of eating a snail, is a little hard to figure out. Anyone who, through necessity or curiosity, has tasted them, pronounces them excellent, and the writer can testify that they are equal in flavor and texture to any other mollusk. On the open sea coast, this species is a robust, thick-shelled snail, over an inch long. In protected bodies of water, such as Long Island Sound, they seldom attain a length of more than three-fourths of an inch and are correspondingly thinner-walled. The writer has found these periwinkles active in tide pools in January, with the air temperature close to the zero mark.

LITTORINA OBTUSATA Linne (Smooth Periwinkle)

p. 156

The smooth periwinkle is nearly one-half inch in height and very variable in color. The shell is quite stout and globular, smooth and shining, with very faint revolving lines. There are four whorls, the last one very large and the others scarcely rising above it. The aperture is nearly circular, the outer lip thin and sharp, the operculum horny, and the umbilicus lacking. This is a very common species, inhabiting positions exposed to the sea, on rocks and seaweeds. Many specimens are bright yellow, while others may be orange, whitish, or reddish brown. Some, especially the juveniles, are boldly banded. Formerly listed as *Littorina palliata* Say, this little periwinkle is found from Maine to New Jersey.

LITTORINA SAXATILIS Olivi (Rough Periwinkle)

p. 156

The rough periwinkle averages less than one-half inch in height, and is yellowish gray or ash-colored in hue. The shell is ovate, strong, and coarse. There are four or five convex whorls, well defined by the sutures, forming a moderately elevated spire. The surface is marked with very perceptible lines of growth, and by revolving grooves. The aperture is oval, and the operculum is horny. Formerly known as *Littorina rudis* Maton, this is a very common shell from New England north, where it is found on the rocks, often above the high tide level. The young shells are smooth and variously mottled and spotted with yellow and black.

LITTORINA IRRORATA Say (Gulf Periwinkle) p. 156

This snail is about one inch high, and its color is soiled white, with spiral lines of chestnut-brown dots. The shell is heavy and robust, and sharply pointed. There are five whorls, decorated with fine spiral ridges. The outer lip is stout, tapering rapidly to a thin edge. The aperture is oval. The regular rows of brownish dots give the shell a chestnut color.

This snail replaces the common periwinkle, *littorea*, in the south, being very abundant in the Gulf of Mexico, and ranging north to New Jersey. Its station in life is between the tides, where it may be found on marine vegetation. In the past it existed in the north, and fossil specimens (Post-Pleistocene) are commonly found in the peat beds on the Connecticut shore of Long Island Sound.

LITTORINA ANGULIFERA Lamarck (Southern Periwinkle)

p. 156

This shell is a little more than one inch in height, and it is somewhat variable in color. It may be gray, reddish, yellow, or purplish, with dark, oblique markings. The shell is rather thin for a member of this genus. There are six or seven whorls, terminating in a sharp apex. The sutures are slightly channeled.

There is a sculpture of minute spiral lines. The outer lip bears a central groove in the lower part, and the aperture is large and oval. The operculum is horny. Like the other periwinkles, this species, once known as *L. scabra* Orbigny, may be collected without difficulty, even at high tide. They may be seen clustered on wharf pilings and the roots and leaves of mangroves, in Florida and the West Indies.

LITTORINA ZICZAC Gmelin   (Zebra Periwinkle)   p. 156
The little zebra periwinkle is about one-half inch in height, and its color is whitish, with many wavy stripes of dark brown or black. The shell is sturdy, with a well-defined keel near the base of the body whorl. There is a sculpture of very fine and widely spaced spiral grooves. The aperture is oval and small, and the operculum is horny. This species is abundant throughout southern Florida and the West Indies. It may be seen by the score on rocks at low tide. Like the other periwinkles, it is an herbivorous snail and appears to live almost exclusively upon the algæ that cover the rocks and stones between the tides.

Genus *Tectarius* Valenciennes 1833

2 Species

TECTARIUS MURICATUS Linne   (Knobby Top Shell)
p. 156
The knobby top is a yellowish-gray gastropod, about three-fourths of an inch in height. The shell is sturdy and top-shaped, with about seven well rounded whorls, slightly shouldered above, and a sharp apex. The surface is decorated with revolving rows of beads, or nodules, about five rows to a volution. The aperture is moderately small and oval. The outer lip is somewhat thickened, and there is a groove down the columella toward the base. This, too, is a rock snail, with habits much like the *Littorinas*. It may be found all the way from the water's edge to several feet above the high-tide mark. It is able to survive for long periods out of water; indeed, Doctor Pilsbry records a specimen which revived after being isolated in a cabinet for a year! The knobby top shell ranges from southern Florida to the West Indies.

Genus *Echinella* Swainson 1840

1 Species

ECHINELLA NODULOSA Gmelin   p. 156
This shell is just under one inch in height, and its color is mottled gray and black. There are about eight whorls, decorated with revolving rows of pointed knobs, larger and more pronounced along the sutures, which are generally lighter colored than the rest of the shell. The aperture is nearly round, the

outer lip moderately thick, and there is no umbilicus. This is a fairly common shell on the Florida Keys, where it congregates on the rocks near the high water line.

Family Lacunidæ

KNOWN AS "chink" shells, these snails are conical, stout, and thin-shelled. The aperture is half-moon-shaped, and the distinguishing character is a lengthened groove, or chink, alongside the columella.

Genus *Lacuna* Turton 1827

7 Species and varieties

LACUNA VINCTA Montagu   (Little Chink Shell)   p. 136
The little chink shell is about one-third of an inch high, and its color is dingy white, commonly banded with purple. The shell is thin and conical, with about five rather stoutish whorls which are separated by moderately deep sutures. The apex is pointed. The surface bears minute lines of growth, and the aperture is semilunar, with the outer lip thin and sharp. The inner lip is flattened, and excavated by a smooth, elongate groove, terminating in a tiny umbilicus. This little snail is easily recognized by its peculiar umbilicus, which forms a prominent groove beside the columella. The snail is found on seaweeds and other marine growths. After storms they can often be collected in quantities from material that has been washed ashore. The little chink shell occurs from Labrador to New Jersey.

Family Turritellidæ

GREATLY ELONGATED, many-whorled shells, generally turreted. A large family, living chiefly in Pacific waters. Relatively few representatives are present on our Atlantic coast.

Genus *Turritella* Lamarck 1799

4 Species

TURRITELLA EXOLETA Linne   (Turret Shell)   pp. 29, 156
The turret shell gets to be about three inches in height, and is brown and white in color. The shell is slender and solid, regularly conic, and has a long spire and a sharp apex. There are from sixteen to eighteen whorls, the upper few of which are not occupied by the animal but are divided by a septum at each

half-turn. The sutures are distinct. The surface bears fine revolving lines. The aperture is round, the lip thin and sharp, and the operculum corneous. This is a strikingly beautiful shell, trim and graceful in outline. The color is often creamy white, splashed with chocolate brown. It makes its home in moderately deep water, in southern Florida and the West Indies.

TURRITELLA VARIEGATA Linne   (Variegated Turret)
p. 201
This is a gracefully elongate shell some two inches in height. There are about ten whorls, flattened a little at their centers, and the sutures are deeply impressed. Each volution is divided into an upper and a lower portion by a rather shallow groove, or channel. The color is whitish or purplish, with reddish longitudinal streaks. This fine shell occurs in Texas and the West Indies.

Genus *Turritellopsis* Sars 1878

2 Species

TURRITELLOPSIS ACICULA Stimpson   p. 156
This is a thin, brownish-white shell, nearly one-half inch tall. There are about ten well rounded whorls, spirally ribbed, and with weaker vertical striations. The aperture is round, and the outer lip is thin and plain. This species is found from Labrador to Cape Cod.

Genus *Tachyrhynchus* Morch 1868

2 Species

TACHYRHYNCHUS EROSA Couthouy   p. 156
This is a pale-brown to dark-brown shell, about one inch in height. It is an elongate, high-spired shell, with about ten flattish whorls, each deeply grooved with five blunt furrows which give the surface a spiral ornamentation. The aperture is almost round, and the operculum is horny. Occurring in fairly deep water from Labrador to Massachusetts, this species is often taken from the stomachs of fishes. The apex is frequently badly eroded or broken in the majority of adult shells.

Family Vermetidæ

SHELL TUBULAR, extended, and irregular in growth. Generally spiral when very young. The animal is wormlike, and the shells are usually erroneously called "worm tubes." They often grow on one another in tangled masses, usually attached to stones or imbedded in sponges or corals.

Genus *Vermetus* Daudin 1800

4 Species

VERMETUS IRREGULARIS Orbigny   p. 141
Individually, this is a loosely coiled, twisted, elongate, tubelike shell, growing in almost any contorted manner, but it is usually intertwined with others of its own kind to form an intricate mass. The very juvenile tip shows a spiral beginning to the shell, which is ribbed longitudinally and is generally considerably roughened by wrinkles and lines of growth. The color is reddish brown on the outside; smooth, white, and polished on the inside. This wormlike, colonial-living gastropod may be found in Florida where it sometimes forms reefs with the accumulation of its twisted shells.

Genus *Serpulorbis* Sassi 1827

1 Species

SERPULORBIS DECUSSATUS Gmelin   p. 141
This is another wormlike shell, twisted and irregular in growth, but it is more solitary and does not live as commonly in tangled masses. The shell is quite strong, and yellowish in color, often marked with brown. It bears longitudinal ridges and circular wrinkles. This shell may be found growing on stones and other shells, in moderately shallow water, from North Carolina to the West Indies.

Genus *Vermicularia* Lamarck 1799

1 Species

VERMICULARIA SPIRATA Philippi   (Worm Tube)

p. 156
This is a yellowish-brown shell, from three to six inches in length, occasionally longer. In its youthful stages the shell is tightly coiled and looks like a small *Turritella*, but as it grows older it becomes free and wanders off in an irregular and seemingly aimless fashion. The tubelike shell is marked with longitudinal, angled keels. The aperture is round, and is closed by a circular, horny operculum that is concave on the outside.

This curious creature also appears at first glance more like a worm than a mollusk, but it is a true gastropod nevertheless. The body is greatly elongated, and the head is provided with tentacles, eyes, and a toothed tongue (radula) that is thoroughly snail-like. Several individuals are generally found growing together in an intricate, tangled mass. This species occurs in shallow water all along the Atlantic coast, as well as in the Gulf of Mexico.

Family Cæcidæ

VERY MINUTE, tubular shells, spiral in the beginning but soon becoming merely cylindrical. The spiral or coiled portion is nearly always lost. The family contains but a single genus, distributed in warm and temperate seas.

Genus *Cæcum* Fleming 1817

14 Species

CÆCUM PULCHELLUM Stimpson   p. 141
This shell is only about one-tenth of an inch in length, and yellowish brown in color. The tiny shell is tubular, slightly curved, and composed of a series of regular, rounded rings. This shell would be passed up by most searchers at the seashore, for it is so small that it could very easily be missed. Under a lens, however, it proves to be an attractive, unusual shell. Its range is from New Hampshire to the Florida Keys.

Family Trichotropidæ

SMALL, THIN, elongate shells, with horny opercula and a hairy periostracum. The angles are keeled, and there is a small umbilicus.

Genus *Trichotropis* Broderip & Sowerby

5 Species

TRICHOTROPIS BOREALIS Couthouy   p. 188
This snail is about one-half inch in height and brown in color. The shell is quite solid, consisting of about four whorls, with deeply channeled sutures. The body whorl is relatively large, and encircled by two prominent rounded ribs or keels, as well as two or three less conspicuous ones. There is a yellowish periostracum, which rises like a bristly fringe along the keels and along the growth lines. The aperture is broad and rounded behind and somewhat narrowed and pointed in front. This interesting snail occurs from Labrador to Massachusetts. It is rather common in the stomachs of fishes taken off the New England coast.

Family Planaxidæ

SMALL, BROWNISH snails, living under stones in shallow water and characterized by a spirally grooved ornamentation. The aperture is oval, well-notched below. Common in warm seas.

Genus *Planaxis* Lamarck 1822

2 Species

PLANAXIS LINEATUS Da Costa   p. 188
This shell is about one-quarter of an inch high, and its color is yellowish brown, generally with spiral brownish bands. The shell is quite solid and well inflated, with three or four whorls, decorated with evenly spaced spiral grooves. The aperture is oval, with a notch at the lower margin. The outer lip is relatively heavy, sometimes slightly deflected above. This is a somewhat variable species, although the majority are conspicuously banded. They occur, often in large numbers, under stones and driftwood, in Florida and the West Indies.

Family Modulidæ

FLATTISH, TOP-SHAPED shells, the whorls grooved and tuberculated. There is a small, narrow umbilicus. The columella ends below in a sharp tooth. Found in warm seas.

Genus *Modulus* Gray 1842

1 Species

MODULUS MODULUS Linne   (Atlantic Modulus) p. 156
This is a knobby little snail, about one-half inch in diameter. Its color is yellowish white, spotted and marked with brown, and there is a thin, gray-brown periostracum. The spire is quite low, there are three or four whorls, and the body whorl is large, with sloping shoulders. The periphery is sharply keeled. The sculpture consists of low revolving ridges, and stout vertical ribs separated by deep grooves. The aperture is nearly round, and the outer lip is thin and crenulate. The operculum is corneous. This snail lives in shallow water, generally in more or less protected bays and lagoons. It may be observed, sometimes in large numbers, on seaweeds and various marine growths, from North Carolina to Florida. A somewhat more knobby form used to be known as variety *floridanus* Conrad, but now all of the

specimens found in Florida are considered as one single, more or less variable, species.

Family Triphoridæ

AN INTERESTING family of very small, elongate, left-handed (sinistral) snails. There are numerous whorls. Widely distributed in warm seas.

Genus *Triphora* Deshayes 1824
31 Species and varieties

TRIPHORA PERVERSA NIGROCINCTA Adams   p. 156
This shell is less than one-half inch in height, and its color is pale brown. The shell is elongate, with about fifteen whorls, the sutures slightly excavated. The surface bears revolving rows of beadlike tubercles, four rows on the body whorl. The spiral turns to the left, instead of to the right as it does in most of the gastropods. This variety may be found in shallow water, from Massachusetts to Florida.

Family Cerithiopsidæ

SHELLS SMALL, cylindrical, and many-whorled, with a nearly straight canal. Ornamented with a spiral sculpture. Usually found in moderately deep water, mostly in temperate seas.

Genus *Cerithiopsis* Forbes & Hanley 1849
28 Species and varieties

CERITHIOPSIS CRYSTALLINA Dall   p. 156
This is a slender little shell, not quite one-half inch in height. There are about eight whorls, with distinct sutures. The shell is pure white and is decorated with revolving rows of tiny tubercles. The lower portion of the body whorl bears spiral grooves. Under a magnifying glass this will prove to be a strikingly beautiful shell. It is to be found in rather deep water, in the Gulf of Mexico.

CERITHIOPSIS SUBULATA Montagu   p. 156
Just under one inch in height, this shell has from ten to twelve flat whorls. The sutures are indistinct. The shell is decorated with revolving rows of tiny beads. The base of the shell is

THE LITTORINAS AND MISCELLANEOUS SNAILS

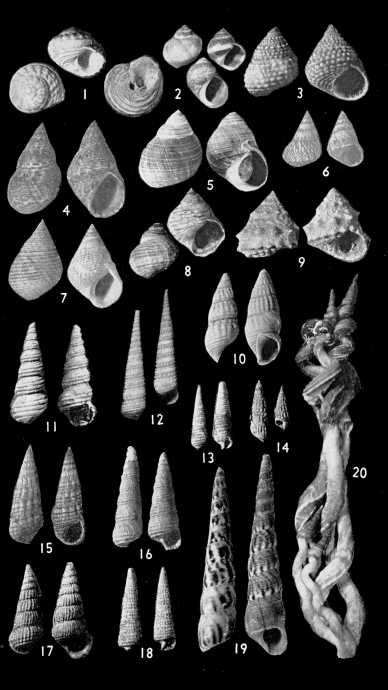

Plate 32   157

THE CONCH SHELLS AND THE HORN SHELLS

1. *Cerithium eburneum* × 1, 2 views: **BEADED HORN SHELL**
   p. 159
       See text.

2. *Batillaria minima* × 1, 2 views: **BLACK HORN SHELL**
   p. 162
       Small; black, often with pale bands.

3. *Cerithium muscarum* × 1, 2 views: **DOTTED HORN SHELL**
   p. 159
       See text.

4. *Cerithium floridanum* × 1, 2 views: **FLORIDA HORN SHELL**
   p. 158
       See text.

5. *Strombus alatus* × 1, 1 view: **FLORIDA CONCH**   p. 164
   Strong and solid; expanded lip.

6. *Strombus alatus*, juvenile × 1

7. Clawlike operculum of *Strombus*

8. *Aporrhais occidentalis* × 1, 2 views: **DUCK FOOT**   p. 163
   Grayish white; outer lip flaring.

9. *Strombus gigas* × ¼, 1 view: **QUEEN CONCH**   p. 165
   Heavy and ponderous; pink inside.

10. *Tonna maculosa* × ⅔, 1 view: **PARTRIDGE SHELL** p. 176
    Large but light; chestnut crescentic markings.

smooth, with cordlike ridges. The aperture is small, the outer lip somewhat thickened, and the inner lip is slightly twisted. The color is pale brown. This interesting univalve occurs from Massachusetts to the West Indies.

Genus *Seila* Adams 1861

2 Species

SEILA ADAMSII Lea   p. 156
Formerly known as *Cerithiopsis terebralis* Adams, this is another small snail, elongate, brown in color, and made up of many whorls. Its average height is just about one-half inch, and there may be as many as a dozen whorls. The sculpture consists of three or four elevated, revolving ridges on each volution, with fine vertical lines between the ridges. The base is short, and the aperture is rather small. Occurring from central New England to Florida, this diminutive gastropod inhabits rather shallow water, and empty shells are often found in the drift along shore.

Genus *Cerithiella* Verrill 1882

2 Species

CERITHIELLA WHITEAVESII Verrill   (Whiteaves' Cerithiella)
                                                                        p. 204
This is a small elongate shell, composed of about eight rather sloping whorls. The apex is somewhat blunt, the sutures well defined, and the aperture small. The volutions are sculptured with strong vertical ribs, made wavy by a double row of spiral knobs. The outer lip is plain, and the inner lip is twisted. The color is brown. This little fellow lives from Massachusetts to the Gulf of St. Lawrence.

Family Cerithiidæ

A LARGE FAMILY of generally elongate, many-whorled shells, living in moderately shallow water, mostly upon grasses and seaweeds, in tropical and semitropical seas. The aperture is small and oblique, and there is a short anterior canal. They are commonly known as "horn shells."

Genus *Cerithium* Bruguière 1789

28 Species and varieties

CERITHIUM FLORIDANUM March   (Florida Horn Shell)
                                                                        p. 157
The Florida horn shell is about one and one-fourth inches high, and white in color, with a spiral pattern of brown. The shell is

elongate, with about ten whorls and indistinct sutures. The apex is sharp. There is a sculpture of elevated, nodular ribs, sharply angled at the edge of the volutions, and many unequal ridges and fine lines spiralling over the entire shell. The aperture is small, oval, and oblique. This is a graceful shell, quite decorative, found from North Carolina to southern Florida, where it may be seen crawling over the grassy rocks and weeds in tide pools, and in shallow water generally, feeding upon algæ and various aquatic plants.

CERITHIUM MUSCARUM Say   (Dotted Horn Shell)

p. 157

This shell is not quite one inch in height. Its color is white, or grayish white, with small chestnut dots. The shape is elongate-conical, with an acute apex. There are nine or ten whorls, the sutures rather distinct. There is an ornamentation of distant, prominent, longitudinal ribs, about eleven on the body whorl, crossed by impressed spiral lines. The aperture is oblique and oval, and there is a short, recurved canal. The regularly arranged brown dots on a whitish background give this species a neat and delicate appearance. Like most of the horn shells, it prefers grassy bottoms in shallow water, where it may be observed in large numbers clinging to the aquatic plants. Its range is from central Florida to the West Indies.

CERITHIUM LITERATUM Born   (Lettered Horn Shell)

p. 29

The lettered horn shell is just under one inch in height as a rule, sometimes a little larger. Its color varies from white to gray, irregularly marked with brownish black. The shell is rather stoutly conical, the apex sharply pointed. There are about seven whorls, with indistinct sutures. Two rows of prominent knobs encircle each volution, those on the shoulders largest, and between the knobs are spiral beaded lines. The outer lip is thickened, the aperture is moderately large and oval and the canal is short and only partially recurved. This species is somewhat stouter than most of its group. It is a conspicuous shell, with a black-and-white pattern that is easily seen as the mollusk crawls about over the rocks and seaweeds in shallow water. It occurs from central Florida south, and the searcher will most likely find a few specimens in every tide pool throughout its range.

CERITHIUM EBURNEUM Bruguière   (Beaded Horn Shell)

p. 157

This species is about one-half inch high, and its color is variable, ranging from pure white to deep brown. The shell is elongate, with the spire tapering to a pointed apex. There are six or seven whorls, the sutures distinct, and a sculpture of revolving bands of small tubercles, giving the surface a beaded appear-

THE HELMET SHELLS

1. *Cassis madagascarensis* × ⅓, 1 view: **YELLOW HELMET**
   p. 171
   Large and heavy; oval from below.

2. *Tonna galea* × ⅔, 1 view: **FLORIDA CASK SHELL**   p. 176
   Body whorl swollen; light but sturdy.

3. *Cassis flammea* × ½, 1 view: **FLAME HELMET**   p. 174
   See text.

4. *Phalium granulatum* × ⅔, 2 views: **SCOTCH BONNET**
   p. 175
   Regularly spaced squarish dots.

5. *Cassis tuberosa* × ⅓, 2 views: **SARDONYX HELMET** p. 174
   Large and heavy; triangular from below.

Plate 34   161

THE COWRIES AND THE ROCK SNAILS

1. *Simnia uniplicata* × 2, 2 views   p. 166
   Very slender; often purplish.

2. *Cypræa exanthema* × ½, 2 views: **MEASLED COWRY**
                                                        p. 167
   Brown teeth; heavily spotted; polished.

3. *Cypræa* cut to show early whorls

4. *Trivia quadripunctata* × 1, 2 views: **FOUR–SPOTTED COF-
   FEE BEAN**   p. 170
   Four spots on dorsal surface.

5. *Trivia pediculus* × 1, 2 views: **COFFEE BEAN**   p. 170
   Six spots on dorsal surface.

6. *Muricidea hexagona* × 1, 1 view   p. 183
   Surface rather spiny.

7. *Urosalpinx tampænsis* × 1, 2 views: **TAMPA DRILL** p. 184
   Shoulders angled; strongly mottled.

8. *Murex fulvescens* × 1, 1 view: **TAWNY MUREX**   p. 181
   Large and stout; canal nearly closed.

9. *Murex brevifrons* × 1, 2 views: **SHORT–FROND MUREX**
                                                        p. 182
   Large; frondose, canal long.

10. *Muricidea multangula* × 1, 2 views   p. 183
    White, spotted with brown; aperture rosy.

11. *Murex recurvirostris rubidus* × 1, 2 views: **RED   ROCK
    SHELL**   p. 181
    See text.

12. *Cymatium chlorostomum* × 1, 2 views: **BROWN–MOUTHED
    TRITON**   p. 178
    Surface with network pattern.

ance. The aperture is oval, and the canal is very short. The beaded horn shell is a variable species, occurring in southern Florida. Its granulated appearance and lack of strong longitudinal ribs or folds serve as identifying characters. It can generally be found with little trouble on tidal flats when the tide is out.

Genus *Batillaria* Benson 1842

1 Species

BATILLARIA MINIMA Gmelin   (Black Horn Shell)

p. 157

The black horn shell is about one-half inch tall, and black or deep brown in color, frequently with a paler band encircling the shell just below the suture. The snail is gracefully elongate, with from six to eight whorls. The sutures are well defined, and the apex is sharp. There is a sculpture of low longitudinal ribs, broken by unequally nodular spiral ridges. The aperture is oval, and the canal is short and turned to the left. This is probably one of the commonest shells to be found on muddy shores in southern Florida. It is an active mollusk, leaving a network of tangled trails as it wanders about over the surface of the mud. It is an extremely variable shell in its markings, and a jet-black variety is known as *Batillaria minima nigrescens* Menke.

Genus *Cerithidea* Swainson 1840

6 Species

CERITHIDEA TURRITA Stearns   (Turreted Horn Shell)

p. 156

The turreted horn shell is about one-half inch high, and pale buff in color, with spiral bands of brown. The shell is elongate, tapering regularly to a sharp apex. There are about ten rather convex whorls, sculptured with crowded longitudinal ribs that are crossed by fine spiral lines. The aperture is oval, with a small notch at the upper margin, and a short canal, recurved, at the base. This is a pretty little shell, found rather commonly on the west coast of Florida. It appears to prefer shallow water, and is most often found in grassy situations, where it can generally be collected on the aquatic vegetation close to shore.

CERITHIDEA SCALARIFORMIS Say   (Ribbed Horn Shell)

p. 156

This is a larger species, about one inch in height, and its color varies from pale gray to buffy brown. The shell is high-spired, with about nine well rounded whorls, the sutures very distinct. Each volution is decorated with closely spaced longitudinal ribs. The base of the last whorl bears spiral riblets. The aperture is large and circular, the outer lip partially reflected, and the operculum is horny. This decorative little horn shell may be found from Georgia to Texas. It lives near the high-water mark on grassy bottoms, and is quite abundant on the Florida Keys.

Genus *Bittium* Leach 1847

5 Species

BITTIUM ALTERNATUM Say   p. 156
This is a diminutive snail, about one-fifth of an inch in height. Its color is bluish black, or slate color, the lower whorls some-times paler. The shell is elongate and somewhat turreted, with from six to eight whorls which are covered with a granular network of elevated spiral lines crossed by rounded folds or ribs. The aperture is obliquely rounded, the outer lip is sharp, and the canal is a mere notch or fissure. The operculum is horny. This little gastropod is found clinging to seaweeds and submerged objects just below the low water level. The young are reddish brown, and sometimes occur in such numbers that the sand appears to be alive with them. This species is found from Massachusetts to New Jersey.

Family Aporrhaidæ

STRONG AND SOLID shells, with a high spire and a long and narrow aperture. The outer lip is greatly thickened and flaring, forming a winglike expansion. The best-known example of this group is the "pelican's foot" of European waters.

Genus *Aporrhais* Dillwyn 1823

1 Species

APORRHAIS OCCIDENTALIS Beck   (Duck Foot) p. 157
The duck foot is white or grayish white in color, and about two inches in height. The shell is thick and conical, with from eight to ten whorls, each decorated with numerous smooth, rounded, crescent-shaped folds. On the body whorl there are about twenty of these folds. The shell is further ornamented by closely spaced revolving lines. The aperture is semilunar, with the outer lip expanded into a wide, three-cornered wing. This odd-appearing gastropod is seldom found on the beach, but fishermen now and then bring up specimens in their nets, and they are occasionally taken from the stomachs of fishes. The peculiar winglike lip makes this shell easy to identify. It is a

northern species, inhabiting rather deep water from the Gulf of St. Lawrence to Cape Cod.

Family Strombidæ

AN INTERESTING FAMILY of very active, carnivorous snails, widely distributed in warm seas. The shells are thick and solid, with a greatly enlarged body whorl. The aperture is long and narrow, with a notch at each end, and the outer lip, in adults, is usually thickened and expanded. The operculum is clawlike, and fails to close the aperture.

Genus *Strombus* Linne 1758

9 Species and varieties

STROMBUS PUGILIS Linne   (Fighting Conch)   p. 104
The fighting conch is a robust fellow, from three to five inches in height. The color is deep yellowish brown, commonly with a paler band midway of the body whorl. The early whorls are sculptured with revolving ribs, and there are prominent spines or knobs on the shoulders of the later whorls. The inside of the aperture is deep orange or purple. The shell is large and solid, with a short spire of six or seven volutions, and a much enlarged body whorl. The aperture is quite narrow, and the outer lip is widely flaring and deeply notched below. The operculum is slender and clawlike, and does not completely close the aperture.

This snail is very common in the West Indies and is now and then seen in southern Florida, although it is not common there. A closely related species, to be described next, appears to take its place on our shores. The fighting conch is an active mollusk, and it makes good use of its clawlike operculum in moving about when stranded by the tide. The claw is hooked into the sand and the shell is then raised high in the air, to topple over in a most undignified way, the snail thus progressing in a rather awkward but effective manner.

STROMBUS ALATUS Gmelin   (Florida Conch)   p. 157
The Florida conch is also about three to five inches in height. Its color is yellowish brown, clouded and sometimes striped with orange and purple. The shell is quite solid, with a spire of about seven whorls. The early volutions are decorated with revolving ribs, and the shoulders of the later whorls may or may not bear blunt spines.

There has been much confusion regarding the identity of this and the preceding one, *S. pugilis*. Since the original figure of *alatus* showed a specimen without shoulder spines, it is commonly believed that the spineless form is *alatus*, and the form with well developed spines *pugilis*. However, *alatus* also has spines in many cases. The present form (*alatus*) is very abundant in Florida, and is usually quite mottled, and its shoulder spines, when present, are confined to the last whorl, while the typical *pugilis* is a West Indian form, with prominent spines on the last two whorls. Furthermore, the later is a somewhat heavier shell, and is generally of a more solid, orange color.

The fighting conchs are carrion-eaters, pugnacious, as the name implies, and may be found in numbers feeding upon dead fish in shallow water. Juvenile, or half-grown individuals, do not have the flaring lip, and look considerably like cone shells (*Conus*).

STROMBUS RANINUS Gmelin. (Hawk Wing) p. 104
This snail is known popularly as the "hawk-wing." It is from four to six inches in length, and yellowish white in color, streaked and blotched with brown and black. The aperture is pinkish. The shell is large, sturdy, porcellaneous, and has about eight whorls and a well developed spire. The shoulders of the body whorl bear small knobs, and there is a single large knob upon the back, with a smaller one between it and the margin. The outer lip is thick and flaring, and deeply notched near the canal. The operculum is clawlike. The hawk-wing is found in southern Florida and the West Indies. Like the others of its genus, it is a scavenger, feeding upon dead organisms encountered as it crawls about in moderately shallow water.

STROMBUS COSTATUS Gmelin (Ribbed Stromb)
p. 201
This is a large and solid shell, three to four inches in length. There are about ten whorls, producing a fairly tall spire, and a greatly expanded body whorl. There is a series of large, blunt nodes on the last whorl, with a series of smaller nodes spiralling up the spire. There are also a number of rather distinct horizontal corrugations on the body whorl. The aperture is long and narrow, with the outer lip expanded and very much thickened in old specimens. The color is yellowish white, more or less mottled, the interior pure white with a noticeable aluminum-like glaze on the columella and outer lip. This species may be found from Florida to South America.

STROMBUS GIGAS Linne (Queen Conch) p. 157
The queen conch is from eight to twelve inches in length. Its color is yellowish buff, the interior rose pink. The shell is heavy and solid, with a short, conical spire. The whorls bear blunt nodes on the shoulders. The aperture is narrow and

elongate, channeled at both ends, and the outer lip is thickened and greatly flaring when the mollusk is fully grown. The horny operculum is clawlike.

This is the shell that for generations has been used as a door-stop, or for decorating the borders of flower beds, by families of seafaring men. It is one of the largest and heaviest gastropods, individuals sometimes weighing more than five pounds. It occurs from Florida to the West Indies. Feeding largely upon carrion, the queen conch is a sprightly and energetic snail, well able to overpower and consume living bivalves. It is a commercial shell, and large numbers are exported from the Bahamas for cutting into cameos, the scrap material being ground to powder for manufacturing porcelain. Its flesh is eaten in the West Indies, the aborigines formerly making scrapers, chisels, and various other tools from its shell. Semiprecious pearls are occasionally found within the mantle fold.

STROMBUS GALLUS Linne   (Cock Stromb)   p. 104
This is a smaller species, from four to seven inches long. Its color is mottled brown, white, and orange. The shell is strong and solid, with a sharp spire of about seven whorls. There is a sculpture of strong revolving ribs, and the body whorl bears blunt spines, or nodes, at the shoulders. The outer lip is expanded, and extends above the top of the spire. This is a West Indian shell, occurring occasionally in southern Florida. It is notable for the development of its flaring, winklike lip.

Family Ovulidæ

SHELLS USUALLY long and slender, with a straight aperture, notched at each end. Occurring in warm seas, the mollusks attach themselves to sea-fans and various marine growths.

Genus *Simnia* Risso 1826

2 Species

SIMNIA UNIPLICATA Sowerby   p. 161
This shell averages about three-fourths of an inch in length, and is generally some shade of pink or purple in color, although it may be white or yellowish. The shell is thin, elongate, somewhat cylindrical, and for an aperture it has a narrow slit running the full length of the shell. The ends are bluntly rounded. The surface is smooth and polished. This species may be found from North Carolina to the West Indies. It lives attached to the stems of gorgonias or sea-fans, and is usually colored to harmonize with its surroundings.

Genus *Cyphoma* Bolten 1798

2 Species

CYPHOMA GIBBOSA Linne   (Flamingo Tongue)   p. 29
The little flamingo tongue is a shiny white, or creamy white, snail about one inch in length. The shell is long and narrow, solid and durable, with a dorsal ridge, or hump, near the center of the shell and extending squarely across it. The aperture is narrow and runs the length of the lower side, and the inner and outer lips are without teeth. The humpbacked flamingo tongue is generally found living on the sea-fan, or some branching aquatic plant, where it clings tightly to one of the stems. The shell has a natural polish, but dead beach specimens are soon abraded and become dull white. This little gastropod may be found from the Carolinas to the West Indies.

Family Cypræidæ

THESE ARE the cowries, a large family of brightly polished and brilliantly colored shells that have always been great favorites with collectors. The shell is more or less oval and well inflated, the spire usually covered by the body whorl, and the aperture, lined with teeth on both lips, running the full length of the shell. There is no operculum. This is distinctly a tropical family, with scores of richly colored representatives distributed all around the world. Only a few species, however, are hardy enough to live as far north as Florida.

Genus *Cypræa* Linne 1758

4 Species and varieties

CYPRÆA EXANTHEMA Linne   (Measled Cowry) pp. 29, 161
The measled cowry is a handsome fellow from three to four inches long. Its color is purplish brown, with round whitish spots, often in rings. Young specimens are strongly banded with broad streaks of brown, and these streaks often persist in adult shells, occasionally specimens having their backs practically unspotted. The shell is inflated and oval, rather thin, and has the spire completely concealed by the last whorl. The whole surface is highly polished. The aperture is narrow and elongate, extending the full length of the shell on the under side, and is notched at both ends. Both lips are strongly toothed and dark brown in color. The operculum is lacking.
In the cowries the mantle is expanded on each side, forming lobes which meet over the back of the shell, pretty much hiding the shell when the mollusk is crawling. There is usually a line of paler color on the back, showing where the mantle lobes meet.

THE DOVE SHELLS, DYE SHELLS, AND OTHERS

1. *Nassarius ambigua* × 1, 2 views   p. 195
   Stout; fluted whorls.

2. *Nassarius vibex* × 1, 2 views: **MOTTLED DOG WHELK**
                                                              p. 195
   Short and stout; white.

3. *Nassarius trivittata* × 1, 2 views: **LITTLE DOG WHELK**
                                                              p. 194
   White or banded; pimply surface.

4. *Sistrum nodulosum* × 1, 2 views: **RATTLE SHELL**   p. 187
   Dark, with paler rounded knobs.

5. *Nassarius obsoleta* × 1, 2 views: **BASKET SHELL**   p. 194
   Dark purplish black; tip often eroded.

6. *Anachis avara* × 3, 2 views   p. 191
   See text.

7. *Cantharus cancellaria* × 1, 2 views: **CROSS–BARRED SPIN-DLE**
                                                              p. 197
   Network sculpture.

8. *Pyrene mercatoria* × 1, 2 views: **MOTTLED DOVE SHELL**
                                                              p. 190
   Narrow aperture; short and stout.

9. *Pyrene rusticoides* × 1, 2 views: **SPOTTED DOVE SHELL**
                                                              p. 190
   Narrow aperture; more elongate and polished.

10. *Thais floridana* × 1, 1 view: **FLORIDA DYE SHELL** p. 186
    Rough surface; interior orange.

11. *Cantharus tinctus* × 1, 1 view: **MOTTLED SPINDLE** p. 197
    See text.

12. *Thais lapillus* × 1, 1 view: **ROCK PURPLE**   p. 186
    Pale yellow; small but thick and solid.

13. *Mitrella lunata* × 10, 2 views   p. 191
    Minute; orange moonlike markings.

14. *Conus jaspideus* × 1, 1 view: **JASPER CONE**   p. 218
    Prominent spire; banded or mottled.

15. *Conus jaspideus pygmæus* × 1, 1 view: **PIGMY CONE**
                                                              p. 218
    Small; bluish gray, mottled.

16. *Vasum muricatum* × 1, 1 view: **VASE SHELL**   p. 210
    Heavy and rugged; surface rough; top-shaped.

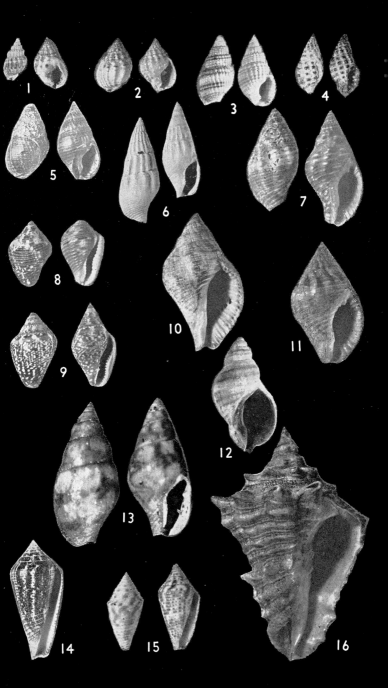

Plate 36   169

THE WHELKS

1. *Buccinum abyssorum* × 1, 1 view   p. 196
   Shoulders angled.

2. *Buccinum totteni* × 1, 2 views: **TOTTEN'S WHELK**   p. 196
   Yellowish brown; surface fairly smooth.

3. *Buccinum tenue* × 1, 1 view   p. 196
   Whorls with vertical folds.

4. *Colus ventricosus* × 1, 1 view: **STOUT COLUS**   p. 203
   Shell swollen; aperture wide.

5. *Buccinum undatum* × 1, 1 view: **WAVED WHELK**   p. 195
   Surface of whorls wavy; aperture yellow inside.

6. *Colus cælatulus* × 1, 1 view   p. 202
   Small; apex blunt; often pinkish.

7. *Colus obesus* × 1, 2 views: **OBESE COLUS**   p. 202
   Small, stout; vertical folds.

8. *Colus spitzbergensis* × 1, 1 view   p. 202
   Large; revolving ribs; aperture circular.

9. *Colus pygmæus* × 1, 1 view: **PIGMY WHELK**   p. 199
   Bluish, with dark green periostracum.

10. *Colus lividus* × 1, 1 view   p. 202
    Elongate; revolving ribs, unequal in size.

The measled cowry is found in moderately shallow water, from North Carolina south. *Cypræa exanthema cervus* Linne is merely a long and stout variety of the above, in which the spots are never ringed. It is usually darker colored, and is believed to be the largest of the cowries, and has the same range as the typical variety.

CYPRÆA SPURCA Linne   (Little Yellow Cowry)   p. 29
The little yellow cowry is about one inch in length. Its color is yellowish white, spotted with yellow. The shell is solid and oval. A tiny spire is present in juveniles but concealed by the last whorl in adults. The aperture is long and narrow, slightly curved, and evenly toothed. The entire shell is highly polished. This little cowry occurs from Florida south and is found as well in European waters. Brilliantly polished shells may sometimes be found among the pebbles on the beach, and living specimens may be seen crawling slowly over the bottom in stony places. This species is not overly common in our country. It is reported as being most abundant in the late spring.

CYPRÆA CINEREA Gmelin   (Gray Cowry)   p. 204
This is a rather plump cowry, about an inch and a half long. The color is grayish brown on the top, with the lower portions shading to lilac. Two paler bands commonly encircle the shell, and the sides are often decorated with black dots and sometimes streaks. The bottom is creamy white, with the apertural teeth rather small. This cowry is found in southern Florida and the West Indies.

Genus *Trivia* Gray 1832
8 Species and varieties

TRIVIA PEDICULUS Linne   (Coffee Bean)   p. 161
This shell is about one-half inch in length, and violet brown in color, with patches of chocolate. The shell is small but solid and subglobular in outline, with the narrow aperture extending the full length of the lower side, and strongly toothed within. The upper side bears an impressed median furrow, with strong radiating ridges which extend around and into the aperture. The surface is not polished.

The shells of this genus resemble those of *Cypræa*, but the animal is quite different. This species is the largest form on our coast, and it may be found rather commonly throughout Florida. The chocolate patches on the back are generally arranged in three pairs, on either side of the median furrow.

TRIVIA QUADRIPUNCTATA Gray   (Four-spotted Coffee Bean)   p. 161
This species is not much over one-fourth of an inch long. Its

color is pale pink, with brownish spots. The small shell is nearly globular, with a strong dorsal furrow. Fine but distinct ridges radiate from this to the elongate aperture. The upper surface bears four dots, staggered, two on each side of the median furrow. This diminutive snail is fatter than the last species. Ranging from Florida to the Bahamas, it feeds upon carrion, and is often found on the bait in lobster and crab traps. The body of this little gastropod is vivid scarlet, and surprisingly large when fully extruded, and one marvels that it can all be packed away in such a tiny shell.

TRIVIA SUFFUSA Gray   (Pink Coffee Bean)   p. 204
This little fellow seldom attains a quarter of an inch in length. The upper surface is sculptured with very fine horizontal lines, divided along the median line by a longitudinal furrow. The lines continue around the shell and into the narrow aperture. The color is rosy pink, sometimes a little darker at the two ends. This species occurs along the Florida west coast, and is very abundant in the West Indies.

Family Cassididæ

THESE ARE mostly large and heavy shells, many of them used for cutting cameos. The shells are thick, and commonly three-cornered when viewed from below. The aperture is long, terminating in front in a recurved canal. The outer lip is generally thickened. These are active, predatory mollusks, living on sandy bottoms in warm seas. They are popularly known as "cameos" or "helmet shells."

Genus *Cassis* Lamarck 1799
4 Species and varieties

CASSIS MADAGASCARENSIS Lamarck   (Yellow Helmet)
p. 160
This is our largest species, attaining a length of about ten inches. It is grayish yellow in color, clouded with brown markings. The aperture is stained dark brown. The shell is heavy and ponderous. There is practically no spire, the greatly enlarged body whorl constituting most of the shell. This body whorl bears three spiral ridges, with blunt knobs. The thickened outer lip has a few well spaced large teeth, while the inner lip bears many riblike smaller teeth, or plications. These are pale buff in color, with the area between them deep brown. The front of the aperture is recurved and folded back on the shell. This large gastropod used to be called *Cassis cameo* Stimpson. It lives on sandy bottoms in shallow water, preying chiefly upon bivalves. It occurs from North Carolina to the West

MISCELLANEOUS SHELLS

1. *Turris albida* × 2, 1 view   p. 219
   Slit in outer lip.

2. Marginal slit of *Turris* × 5

3. *Phos candei* × 2, 2 views   p. 198
   Beaded surface; sometimes lightly banded.

4. *Colubraria lanceolata* × 1, 1 view   p. 187
   Two prominent varices; canal short.

5. *Neptunea despecta tornata* with operculum × 1, 1 view   p. 198
   Spire turreted; revolving elevated ridges.

6. *Latirus infundibulum* × 1, 1 view: **RIDGED LATIRUS** p. 208
   Spiral ridges, reddish in color.

7. *Xancus angulatus* × ½, 1 view: **LAMP SHELL**   p. 209
   Heavy and strong; columella strongly plaited.

8. *Clathrodrillia ostrearum* × 1, 2 views   p. 220
   See text.

9. *Leucozonia cingulifera* × 1, 1 view   p. 209
   Strong and solid; tubercles on shoulders.

Plate 38   173

THE MARGIN SHELLS AND THE OLIVE SHELLS

1. *Marginella apicina* × 1, 2 views: **RIM SHELL**   p. 212
   Small; white; polished.

2. *Marginella avena* × 2, 2 views: **BANDED RIM SHELL**
                                                         p. 213
   Small; polished, with orange bands.

3. *Marginella guttata* × 1, 1 view: **SPOTTED RIM SHELL**
                                                         p. 213
   Larger; mottled, flecked with white.

4. *Olivella mutica* × 2, 2 views   p. 214
   Small; highly polished; no folds on columella.

5. *Oliva reticularis* × 1, 2 views: **NETTED OLIVE**   p. 214
   Reticulated pattern; polished.

6. *Architectonica granulata* × 1, 3 views: **SUNDIAL**   p. 146
   Flatly coiled; umbilicus deep and strongly crenulate.

7. *Busycon caricum eliceans* × 1, 1 view   p. 205
   Greatly swollen canal.

8. Egg ribbon of *Busycon*

9. Operculum of *Murex*

10. *Voluta musica* × 1, 2 views: **MUSIC VOLUTE**   p. 210
    Pattern suggesting bars of music.

Indies. When this shell was first described it was mistakenly thought to come from Madagascar.

CASSIS TUBEROSA Linne   (Sardonyx Helmet)   p. 160
The sardonyx helmet shell is from six to nine inches in length, and its color is buffy or rufus yellow, mottled and blotched with various shades of brown. The shell is much like the last species, but more triangular when viewed from below. In addition to the brown stain between the plications on the inner lip, there is a conspicuous patch of bright chestnut toward the posterior end of the aperture. This species, sometimes called the "black helmet," is a preferred one for cameo-cutting, as there is a very dark coat beneath the outer layer, so that the figure stands out well against the "onyx" background. Most of the other helmet shells yield a cameo with a reddish orange, or pink ground. The range of this species is from North Carolina to the West Indies, and it, too, prefers moderately shallow water and a sandy bottom.

CASSIS FLAMMEA Linne   (Flame Helmet)   p. 160
The flame helmet is from three to five inches in length. It is quite a colorful shell, delicately cream-colored clouded with crescent-shaped markings of brown and black. The shell is solid and heavy, the spire short, and the apex blunt. There are four or five whorls, transversely wrinkled above, the shoulders displaying rows of small nodules. The aperture is elongate, toothed within on both lips, and the outer lip is greatly thickened. The canal is short and recurved. This is a handsome snail, the regular dark brown or black spots on the thickened outer lip contrasting sharply with the delicate shades of the body whorl. It is an active gastropod, found in the Bahamas and West Indies. It is sometimes used for cameo-cutting.

Genus *Cypræcassis* Stutchbury 1837
1 Species

CYPRÆCASSIS TESTICULUS Linne   (Baby Bonnet)
p. 29
The baby bonnet is one of the smaller helmet shells, averaging from two to three inches in length. Its color is pinkish buff, mottled, especially near the margin, with orange-brown. The shell is solid, with a very short spire, and a very large body whorl which comprises most of the shell. There are numerous weak longitudinal ribs, which are crossed on the lower portion of the shell by strong revolving indentations. The aperture is long and narrow, the inner lip is plicate for its entire length, and the outer lip is rolled back and strongly toothed within. The canal is twisted to the left and folded against the shell. There is no operculum. The baby bonnet occurs sparingly from

Cape Hatteras to Cuba, becoming more abundant as one travels
south. It is an active, predacious mollusk, preying chiefly upon
clams encountered as it crawls over the rocky or gravelly bottoms in moderately deep water. In life the mantle covers much
of the shell.

Genus *Phalium* Link 1807

2 Species

PHALIUM GRANULATUM Born   (Scotch Bonnet) p. 160
The Scotch bonnet is pale yellow or whitish, with squarish light-
brown spots distributed over the shell with great regularity. Its
length is from three to four inches. The shell is solid, and composed of about five whorls, the spire short and the sutures indistinct. The surface bears deeply incised transverse lines.
The aperture is large, the outer lip thickened and toothed
within, and the canal is strongly recurved. The Scotch bonnet
is also an active, energetic species, feeding upon marine life,
mainly bivalves, which it finds by grovelling in the mud. The
regular arrangement of its square spots render this snail rather
easy to identify, although they may be faded or almost obliterated in beach specimens. This snail occurs from Cape Hatteras south, and used to be known as *Cassis inflata* Shaw.

Genus *Morum* Bolten 1798

1 Species

MORUM ONISCUS Linne   (Wood Louse)   p. 76
The wood louse is about one and one-half inches high, and its
color is a soiled white, heavily mottled with brown and gray.
The shell is small but very solid, and bluntly cone-shaped, with
a low spire. The surface is covered with rows of warty tubercles,
especially prominent on the shoulders of the last whorl. The
aperture is elongate, with the outer lip considerably thickened,
and toothed within. There is a heavy white callus over the
inner lip. This warty little snail looks very much like one of the
cone shells (*Conus*), but it is more closely related to the last
group. It is found in southern Florida, where it makes its home
among the rocks in fairly deep water. Fresh specimens have a
velvety periostracum which is usually lacking in shells that are
picked up on the beach.

Family Tonnidæ

A SMALL GROUP of large or medium-size gastropods, chiefly of the
tropics. The shell is thin and subglobular, usually with a greatly
swollen body whorl. They are sometimes called "cask shells"
or "wine jars."

Genus *Tonna* Bolten 1798

2 Species

TONNA GALEA Linne   (Florida Cask Shell)   p. 160
The Florida cask shell is an ivory-white snail, blotched and banded with chocolate brown. Its average height is about six inches, but ten-inch specimens have been recorded. The shell is thin but quite sturdy, and greatly inflated, being almost as broad as it is tall. The sutures are somewhat sunken, and the surface is sculptured with widely spaced revolving grooves. The aperture is very large, and the outer lip is thickened at its edge and wrinkled on the inside by the termination of the grooves. The columella is strongly twisted. Juvenile specimens possess a small horny operculum, but this structure is lacking in mature individuals. The cask shell is thin and light, but surprisingly strong for its weight. The animal is very large, and when seen crawling appears far too big for its shell. It occurs in moderately deep water, and may be found from Cape Hatteras to Texas. Very similar shells, now believed by some authorities to belong to this same species, are found in the eastern Atlantic, along the African coast, and in both the Pacific and Indian Oceans.

TONNA MACULOSA Dillwyn   (Partridge Shell)   p. 157
This shell is not as stout as the last one, nor is it quite as large, averaging from five to eight inches in height. Its color is pale to rich brown, heavily mottled with darker shades, and bearing crescent-shaped patches of white, so that the surface does indeed remind one of the plumage of a partridge. The shell is thin and inflated, and the spire is short, with a pointed apex. The aperture is large, and the outer lip is thin and sharp, and slightly crenulate at the edge. The Atlantic coast species, occurring from southern Florida to South America, has long been known as *Tonna perdix*, but the true *perdix* is now believed to be exclusively Indo-Pacific. Our shell has been considered as a variety of the Pacific form, and will be found in some lists as *Tonna perdix pennata*, but it is now regarded as a separate species, entitled to full specific rank. It has been given the earliest available name, *maculosa*, proposed by Dillwyn in 1817.

Genus *Eudolium* Dall 1889

1 Species

EUDOLIUM CROSSEANUM Monterosato   p. 201
This is a moderately large shell, from two to three and one-half inches tall. There are six well rounded whorls, a moderate spire, and a rather blunt apex. The shell is thin, and is decorated with low and widely spaced encircling ribs. The aperture is long and rather wide, and the outer lip is slightly rolled at the

edge in mature shells. The color is pale buffy white. This is a deep-water gastropod, not to be looked for on the beach. It has been dredged in goodly numbers off the New Jersey coast, and is apparently not uncommon in many places. It ranges south to South America, and is found also in the Mediterranean Sea.

Genus *Ficus* Bolten 1798

1 Species

FICUS PAPYRATIA Say (Paper Fig Shell) p. 76
This is a common Florida shell, five or six inches high. Its color is pinkish flesh, with widely spaced pale-brown dots which sometimes are scarcely discernible. The interior is polished orange-brown. The shell is thin and pear-shaped, the body whorl enlarged above, and narrowing to a long, straight canal. The spire is very short, so that the shell appears flat on top. The entire surface is sculptured by fine growth lines, crossed by small revolving, cordlike ribs. The aperture is large, the outer lip thin and sharp, and there is no operculum. The paper fig shell is common on both coasts of Florida, and may be found as far north as Cape Hatteras. In life the mantle folds back over the shell when the animal is crawling, so that only a small portion of the shell is visible. This snail prefers rather deep water, and is usually found crawling rapidly over the sandy bottom, bearing its light shell with apparent ease.

Family Cymatiidæ

THESE ARE more or less decorative shells, rugged and strong, with no more than two varices to a volution. The closely allied *Murex* shells have three. The canal is prominent, and teeth are usually present on the lips. They are distributed in warm and temperate seas.

Genus *Cymatium* Bolten

10 Species

CYMATIUM AQUITILE Reeve (Hairy Triton) p. 77
This shell is from four to six inches in height, and its color is pale brown, banded with gray and white. The shell is strong and solid, with five or six whorls and a rather pointed apex. The surface is cross-hatched with fine lines running in two directions, the shoulders are somewhat nodulous, and there are two robust varices on each volution. The aperture is large, and both the inner and outer lips are wrinkled into small teeth. The canal is short. This is a fine, large snail, formerly called *Cymatium pileare* or *Triton pilearis*. It occurs in southern Florida, the

West Indies, and in the Pacific. Shells of this genus have two strong varices on each whorl, which represent rest periods in shell growth. In the genus *Murex*, to be discussed later, there are three of these varices to a whorl.

CYMATIUM FEMORALE Linne   (Angular Triton)   p. 77
This is a large and spectacular species, attaining a height of about seven inches, with the majority of specimens averaging between three and five. The color is pale yellowish brown, with bands of darker shade. There are seven or eight well-shouldered whorls. The sculpture consists of revolving ribs of two sizes, the larger ones studded with coarse knobs at regular intervals. Two prominent varices are present on each volution. The outer lip is rolled inward at the margin and decorated with large nodules at the ends of the ribs, the top one curving upwards in the direction of the spire. The canal is moderately long, partly reflected, and nearly closed. This is a West Indian shell, occasionally taken in southern Florida.

CYMATIUM CYNOCEPHALUM Lamarck   (Ribbed Triton)   p. 77
This is another strong and rugged gastropod, about three inches in height, and pale yellow in color, sometimes clouded with gray and white. The shell is composed of about five whorls, a low spire, and a long, somewhat curved canal that is nearly closed at its upper end. The surface bears very heavy horizontal or revolving ribs, and there are two robust varices on each whorl. The aperture is large, the inner lip reflected on the last whorl, and the outer lip is strongly crenulate within. This handsome shell is found in Texas, the Florida Keys, and in the West Indies. Due to its strong horizontal ribbing, it is not likely to be mistaken for any other species.

CYMATIUM CHLOROSTOMUM (Lamarck)   (Brown-mouthed Triton)   p. 161
This shell is from two to three inches in height, and a soiled white in color, more or less mottled with brown. The shell is rugged and solid, with about five whorls, a short spire, and two strong varices on each whorl. The surface is divided into squares by the crossing of horizontal and vertical ribs. The outer lip is thick and heavy, with a double row of teeth inside the aperture. The canal is short and curved. This is not a particularly attractive shell. It prefers to live offshore, but its empty shell is often to be found on rocky beaches. It occurs from Florida to the West Indies. The inside of the aperture is generally dark brown, as the shell's popular name implies.

CYMATIUM TUBEROSUM Lamarck   (White-mouthed Triton)   p. 77
This snail averages about two inches in height, and its color is gray, spotted and sometimes banded with brown. The shell is

rugged and strong, with about five whorls and a moderately sharp apex. The surface is sculptured with revolving, nodular ribs, and there are the usual two varices prominently displayed on each volution. The aperture is large, notched at its upper end, and the canal is moderately long and open. This species occurs in the Florida Keys and in the islands of the West Indies. It somewhat resembles the last species, but may be readily separated by its distinctive aperture and canal.

Genus *Distorsio* Bolten 1798

1 Species

DISTORSIO CLATHRATA Lamarck   (Writhing Shell)

p. 77

This is a yellowish-white gastropod, from one to one and one-half inches in height. There is a sculpture of strong revolving and longitudinal ribs, giving the surface a checkered appearance. The aperture is small, and toothed within on both sides. The columella is strongly reflected on the body whorl, and the canal is open and moderately long. This is a rather deep-water snail, living, as a rule, at some little distance offshore. It occurs from Florida to the West Indies, and empty shells are often to be found on the beaches. Living specimens are occasionally picked up following a storm with a strong inshore wind.

Genus *Gyrineum* Link 1807

2 Species

GYRINEUM AFFINE CUBANIANA Orbigny   (Cuban Frog Shell)   p. 204

This is a fine shell, some two inches in length, found in southern Florida and the West Indies. There are about seven whorls, a moderately tall spire, and a pointed apex. The whorls are well inflated, and the shell has a stocky appearance. There are two prominent varices on each whorl, and they are directly opposite each other, so that there is a more or less continuous ridge running up each side of the shell. The surface is further decorated by revolving lines, some of them beaded, and there is a row of larger beads, or nodes, on the shoulders. The aperture is oval, the outer lip thickened and toothed within, and the inner lip is plicate. There is a short anterior canal, and a notch at the upper angle of the aperture. The color is reddish brown, blotched with white. This shell belongs to the group popularly known as "frog shells," formerly comprising the genus *Ranella*

Genus *Charonia* Gistel 1848

1 Variety

CHARONIA TRITONIS NOBILIS Conrad   (Trumpet Shell)   p. 77

The trumpet shell, or giant triton, is from fifteen to eighteen inches long. It is richly variegated with buff, brown, purple, and red, in crescentic patterns that are suggestive of the plumage of pheasants. The aperture is pale orange, the shell is strong and solid, gracefully elongate, the apex bluntly pointed. There are eight or nine whorls, the suture plainly marked, and the body whorl showing distinct shoulders. Widely separated rounded ribs encircle the shell, and each volution bears two rather obscure varices. The inner lip is reflected, and is stained a dark purplish brown, crossed by whitish wrinkles.

The type of this species, *C. tritonis tritonis*, is an Indian Ocean gastropod, found also in the Philippines and in Japan, where empty shells are made into trumpets and teakettles by many of the island tribes. The variety *nobilis*, occurring in southern Florida and in the West Indies, is a stouter and heavier shell, with distinct shoulders on the body whorl when the mollusk is mature. This is one of the showiest of Atlantic shells, not very common in this country but seen frequently in the Caribbean area.

Family Muricidæ

SHELLS THICK and solid, generally more or less spiny. There are three varices to a volution. These are active, carnivorous snails, preferring rocky or gravelly bottoms and moderately shallow water, as a rule. They occur in all seas, but are most abundant in the tropics.

Genus *Murex* Linne 1758

16 Species and varieties

MUREX POMUM Gmelin   (Apple Murex)   p. 77

The apple murex is from one and one-half to two inches in height and yellowish white in color, with broad stripes and mottlings of chocolate brown. The aperture is tinged with rose, and is marked with brownish on the inner lip. The shell is rough and heavy, the spire well developed, and the body whorl is large. The surface is nodular, and sculptured with revolving ridges and fine, cordlike ribs. There are three prominent varices on each whorl, representing rest periods in shell growth. The aperture is large and round, the outer lip is thickened at its edge (forming one of the three varices), and there is a columellar callus reflected over the body whorl. The canal is short, nearly closed, and curved backward from the aperture. The operculum is horny.

This species ranges from North Carolina to Venezuela, those individuals coming from around Jamaica attaining a length of up to four inches. It is found on both rocky and sandy bottoms, where it creeps about in shallow or moderately deep water in search of what it may devour. Normally a carnivorous and predaceous mollusk, the apple murex does not hesitate to feed upon a dead crab or fish when such a delicacy is discovered. Empty shells of this snail are generally plentiful on the Florida beaches.

MUREX FLORIFER Reeve   (Black Lace Murex)   p. 77
The black lace murex is from one to two inches in height, and deep brownish black in color, with the apex and the aperture pink. Juvenile specimens are commonly pale pink all over. The shell is sturdy and ornate, with about seven whorls that are decorated with frondlike spines. There is a sculpture of rounded revolving cords, crossed by three prominent varices, or riblike thickenings. The spines are most striking near the outer lip. The canal is curved backward, nearly closed, and flattened below. This is a well-known shell, found rather commonly on the shore from North Carolina to South America. The mollusk prefers to live in fairly deep water, but it is far from rare close to shore. Young examples, and some of the old well-worn adults as well, are pale grayish pink, but the average normal mature shell is rich brown, with the pink color confined to the inside of the aperture and the tip of the apex.

MUREX CABRITII Barnardi   (Spiny Murex)   p. 77
The spiny murex is a pinkish-buff shell about three inches high. The shell is quite sturdy, and generally spiny, with about six whorls and rather indistinct sutures. The surface bears revolving ridges. The aperture is moderately small, the canal narrow and much elongated, being fully twice as long as the rest of the shell. There are three varices on each whorl, decorated with slender spines, the spines continuing on down the canal. This curious snail occurs from Florida to Texas and the West Indies. In number and length of spines there is considerable variation, and sometimes they are entirely lacking.

MUREX RECURVIROSTRIS RUBIDUS Baker   (Red Rock Shell)   p. 161
This shell is some two inches in height, and mottled gray, brown, and pink in color, sometimes showing weak bands. The shell is small but stocky, sculptured with revolving lines and the customary three varices that may be decorated with short spines. The aperture is round, and the canal is long and nearly closed at the base. The horny operculum is pale yellow. This is a fairly common shell, closely related to the last one, but smaller and without the decorative spines. It occurs in southern Florida and the West Indies.

MUREX FULVESCENS Sowerby   (Tawny Murex)   p. 161
This is the largest member of its genus to be found in the western Atlantic, attaining a length of fully six inches. Its color is

buffy white. The body whorl is very large, giving the shell a somewhat globular appearance. There are seven or eight whorls and a sharp apex, and three varices, each armed with spines that are connected with horizontal ribs. The aperture is large, and the canal is nearly closed. The outer lip and the canal are well supplied with spines. Formerly called *Murex spinicostatus*, this large and handsome gastropod is said to range from the Carolinas to Texas, but it lives in deep water, and is considered a rare species.

MUREX BREVIFRONS Lamarck   (Short-Frond Murex)

p. 161
This is another sizeable fellow, nearly four inches in height. It is grayish white, mottled with yellow and brown. The shell is rough and strong, with about five whorls and rather indistinct sutures. The surface is roughened by transverse wrinkles. The three varices are decorated with spines and hollow fronds, those near the shoulders curving upwards. The aperture is oval, the canal fairly long and partially closed. Ranging from North Carolina to South America, this shell is common in the West Indies, but quite rare in this country.

MUREX ANTILLARUM Hinds   (Antillean Rock Shell)

p. 204
This is a real decorative shell, three inches in length. There are eight or nine whorls, quite convex, a moderate spire, and a very long, partially open canal. There are three equidistant varices, each lined with spines of varying sizes, but with one long spine at the shoulder. These spines continue part way down the canal. Between the varices the shell is sculptured with revolving ribs, some of them more or less beaded. The color is creamy white, varying to brownish purple. The aperture is commonly purplish within. This is a deep water snail, quite abundant throughout the West Indies, and taken occasionally off southern Florida.

Genus *Eupleura* Adams 1853

3 Species and varieties

EUPLEURA CAUDATA Say   (Thick-lipped Drill)   p. 77
This shell is slightly less than one inch in height. Its color varies from reddish brown to bluish white, the margin of the lip frequently bluish. The shell is solid, with about five whorls that are distinctly flattened at the shoulders. The surface bears about eleven stout vertical ribs, of which the one bordering the aperture and one directly opposite, on the left side of the body whorl, are enlarged into stout, knobby ridges. These are crossed by numerous revolving lines. The outer lip is thick and bordered within by raised granules. The canal is short and almost, but not quite, closed. This little snail is often confused with the

oyster drill, *Urosalpinx cinerea* (page 184), to which it is closely related, but the oddly thickened outer lip immediately identifies it. It is an exceptionally active mollusk, found among the rocks at low tide, from Cape Cod to Florida.

Genus *Tritonalia* Fleming 1828
4 Species and varieties

TRITONALIA CELLULOSA Conrad   p. 204
Just under one inch in length, this is a rough and rugged little shell of about six whorls. There are five or six prominent varices, with strong wrinkled lines between them. There is a moderate-sized canal, and the termination of the varices produces a forked appearance at that end. The aperture is small, and commonly purplish within, while the rest of the shell is grayish white. This species lives in fairly shallow water, preferably on a stony bottom, from North Carolina to the Gulf of Mexico.

Genus *Muricidea* Swainson 1840
4 Species

MURICIDEA MULTANGULA Philippi   p. 161
This gastropod averages about one inch in height. The shell is creamy white, flecked with brown, with the aperture rose-pink. There are about six whorls, the body whorl very large, and the sutures distinct. There is a sculpture of seven prominent varices which are interrupted at the sutures. These are crossed by numerous revolving, threadlike lines. The aperture is oval, and the outer lip is thin. The canal is short, and the operculum is corneous. The shell looks somewhat like *Urosalpinx tampænsis*, to be described later, but its shoulders are less sharply angled, and its sculpture is different. It is quite variable in color, and occasionally one finds a solid brown or orange individual. This species occurs from North Carolina to Texas and the West Indies. Freshly taken specimens have a hairy periostracum.

MURICIDEA HEXAGONA Lamarck   p. 161
This shell is from one to two inches in height, and its color is gray or white, tinged with reddish, especially on the apex. The shell is quite solid, and composed of about six whorls, with rather indistinct sutures and a well elevated spire. The surface is somewhat spiny, with about five rows of short spines on the body whorl, becoming less prominent at the base. The canal is moderately long and open, and the outer lip is margined with frond-like protuberances. This species prefers fairly deep water, and is not likely to be found alive unless one dredges out beyond the low-water level. Its empty shell, however, is usually to be seen on the beaches, in southern Florida.

Genus *Urosalpinx* Stimpson 1865

7 Species

UROSALPINX CINEREA Say   (Oyster Drill)   p. 77

The well-known and much despised oyster drill is about one and one-fourth inches high, and a dirty gray in color, with the aperture dark purple. The shell is coarse and quite solid, with five or six rather convex whorls. The surface is sculptured with revolving raised lines, and made wavy by numerous rounded vertical ribs or folds. On the body whorl there are about twelve of these lines, and ten folds. On the spire the lines decrease, leaving only the vertical folds. The aperture is oval, the outer lip thin and sharp, and the canal short. There is a horny operculum.

Next to the starfish, this little snail is the worst enemy the oystermen have to contend with. In some localities, notably the Chesapeake Bay area, their depredations sometimes exceed those of the starfish. Settling upon a young bivalve, the oyster drill quickly bores a neat, round hole through one valve, making expert use of its sandpaper-like radula. Through this perforation the soft parts of the mollusk are sucked, aided by a long proboscis the drill is able to insert into the hole. The oyster drill is one of the commonest univalves along much of our coast, and may be found by the hundreds at low tide on rocky shores from southern Maine to Florida.

UROSALPINX TAMPÆNSIS Conrad   (Tampa Drill)

p. 161

The Tampa drill is about one inch in height, and its color is grayish brown, more or less heavily mottled with white. The shell is rugged, with five or six broadly shouldered whorls and distinct sutures. There are about ten longitudinal ridges to a volution, cut by strongly sculptured revolving cords. The ridges are generally paler in color than the spaces between them. The canal is short and open, and slightly curved. The operculum is corneous, and yellow. This shell occurs in shallow water, along the east coast of Florida, showing a preference for stony bottoms. Like its northern relative, it is destructive to various bivalve mollusks.

Genus *Trophon* Montfort 1810

13 Species and varieties

TROPHON CLATHRATUS Linne   p. 188

This is a graceful little shell, about one-half inch in height. Its color is pale brown. There are about six well-rounded whorls, distinctly shouldered, with the sutures deeply impressed. The stages of growth are plainly marked by an expansion of the lip, producing a number of raised, rather sharp varices which become rounded with age. The aperture is oval, terminating in a curved canal about half the length of the aperture. This is a cold-water snail occurring from Greenland to Maine. Most of the specimens obtained are taken from the stomachs of haddock and other fishes.

TROPHON SCALARIFORMIS Gould   p. 188

This is the largest member of its genus living on our coast. The height is fully one and one-half inches, and the color is whitish or grayish brown. The shell is quite solid, consisting of eight convex whorls, less shouldered than in the preceding species. The surface is covered at close intervals with compressed, whitish ribs or arching plates, the edges sharp when young but smooth and blunt in old specimens. In fully grown shells there are faint revolving lines between the ribs or varices. The aperture is oval and about one-half the length of the shell. The canal is moderately long and slightly curved. This shell occurs in deep water, from Iceland to Massachusetts Bay.

TROPHON TRUNCATUS Strom   (Truncate Trophon)
p. 188

This shell is about one-half inch in height, and brownish gray in color. The shell has about six rounded whorls, with the sutures well defined. There are some fifteen closely set varices to a volution, thin and sharp in young shells, worn and blunt in adults. The canal is fairly long and turned somewhat to the left. This is probably the commonest *Trophon* on our coast, but most of the shells secured, as with so many of our small deep-water forms, come from the stomachs of fishes.

Family Thaisidæ

STOUT SHELLS, with a greatly enlarged body whorl, large aperture, and a short spire. These are predatory gastropods, living close to shore in rocky situations. They are sometimes referred to as "dye shells," as the body secretes a colored fluid that may be green, scarlet, or deep purple. They occur in all seas.

Genus *Purpura* Bruguière 1789
1 Species

PURPURA PATULA Linne   (Wide-mouthed Dye Shell)
p. 105

This is the largest of its group, reaching a length of about three inches. It is dull grayish green or brown in color, the interior often salmon-pink. The shell is rough and solid, the body whorl greatly enlarged and making up most of the shell. The surface bears revolving lines and numerous nodules, very pronounced in partly grown specimens, but often worn and indistinct in old individuals. The aperture is very large, the outer lip is thin and sharp, and there is a polished area on the inner lip. There is a small, horny operculum, too small to close the aperture. This rather common shell occurs from Florida to Mexico. It lives close to shore, and specimens can generally be found clinging to the rocks during the ebb tide. They vary considerably in size, and medium-sized examples are usually more knobby and make better cabinet specimens than fully grown adults.

Genus *Thais* Bolten 1798

6 Species and varieties

THAIS LAPILLUS Linne   (Rock Purple)   pp. 105, 168
The rock purple averages about one inch in height and is quite variable in color, ranging from white to lemon-yellow or orange-yellow, and purplish brown. Some individuals are banded with white or yellow. The shell is thick and solid, with about five whorls, a short spire, and an acute apex. The body whorl is large, and the surface is sculptured with deep revolving furrows and ridges, as well as by many transverse wrinkles. The aperture is oval, the lip arched, and the operculum horny. There is a very short, open canal. There is considerable variation in their shells, aside from their color. Some are larger, some much stouter, and some more angular than others, and in a series of several dozen shells it is possible to find extremes that if studied alone would almost seem to be different species.

This is a familiar shell on northern coasts, on both sides of the Atlantic. It may have migrated to America by way of Iceland and Newfoundland. The rock purple is predaceous, drilling holes through the valves of young oysters and clams, and also commonly feeding upon barnacles. It is very abundant on the rocks at low tide from Rhode Island north. The famous "Tyrian Purple" of the ancients was a dye obtained by crushing the bodies of a Mediterranean snail somewhat like this species.

THAIS FLORIDANA Conrad   (Florida Dye Shell)   p. 168
The Florida dye shell is some two inches high, and grayish in color, more or less clouded with brown. The shell is strong and rugged, with a large body whorl, a pointed apex, and about five whorls with indistinct sutures. The shoulders are sloping, and the surface is decorated with strong revolving ridges. There is a double row of tubercles on the last whorl. The outer lip is thick, and crenulate within, and there is a short canal. This is a common shell on the rocks at low tide, ranging all the way from Cape Hatteras to Texas.

THAIS DELTOIDEA Lamarck   (Banded Dye Shell)

p. 105

This gastropod is only a little more than an inch in height. It is pinkish gray in color, more or less blotched with brown and purple, commonly in the form of broad bands encircling the shell. There are three or four whorls with indistinct sutures. There are three series of nodules on the body whorl, those on the shoulders forming pronounced knobs. The aperture is large, the outer lip simple and thin, and the inner lip bears a fold on its lower part. There is a short canal. The banded dye shell is a stout little fellow, frequently seen on the rocks while the tide is out, although its shell is often so covered with encrustations that it matches the rock and therefore is not easily detected. It occurs from central Florida south.

Genus *Sistrum* Montfort 1810

2 Species

SISTRUM NODULOSUM Adams   (Rattle Shell)   p. 168

This is a bumpy little shell, just over one-half inch in height. Its color is gray or grayish green, with black nodules. The shell is strong and solid, moderately high-spired, and has five or six whorls. The surface is sculptured with spiral rows of prominent nodules, about five rows on the body whorl and usually three on each of the early whorls. The aperture is long and rather narrow, and more or less contracted by teeth on both sides. This species occurs in shallow water, in Florida and the West Indies, but it is not particularly common on our shores, except locally.

Family Colubrariidæ

FUSIFORM SHELLS, with high spires and long apertures, and short, recurved canals. They are distributed in warm and temperate seas.

Genus *Colubraria* Schumacher 1817

2 Species

COLUBRARIA LANCEOLATA Menke   p. 172

This is a decorative little snail, about one inch in height. Its color is pale buff or yellowish white, with scattered spots of orange brown. The shell is rather elongate, with five or six whorls, each volution bearing many finely cut longitudinal lines and two distinct riblike varices. The aperture is moderately small and narrow, and the canal is short. The inner lip some-

MISCELLANEOUS SMALL SHELLS

See text for field marks.

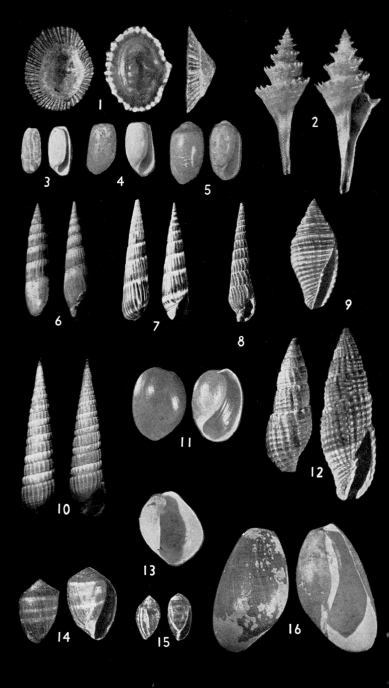

Plate 40   189

THE BUBBLE SHELLS AND OTHERS

1. *Siphonaria alternata* × 1, 3 views   p. 227
   Limpetlike, with bulge on one side.

2. *Ancistrosyrinx radiata* × 2, 2 views   p. 219
   Ornate spines on shoulders.

3. *Cylichna alba* × 2, 2 views   p. 225
   See text.

4. *Retusa pertenuis* × 4, 2 views   p. 224
   See text.

5. *Acteocina canaliculata* × 4, 2 views   p. 223
   See text.

6. *Terebra hastata* × 1, 2 views   p. 216
   Bluntly tapering; pale orange bands.

7. *Terebra cinerea* × 1, 2 views   p. 216
   Whorls with vertical grooves; surface shiny.

8. *Terebra protexta* × 1, 1 view: **BLACK SCREW SHELL**

   p. 215
   Dark brown; slender; spike-like.

9. *Mitra sulcata* × 1, 1 view: **SULCATE MITER**   p. 212
   Aperture narrow; strong plications on columella.

10. *Terebra dislocata* × 1, 2 views: **LITTLE SCREW SHELL**

    p. 215
    Knobby spiral band below suture.

11. *Haminœa elegans* × 1, 2 views: **GLASSY BUBBLE**   p. 226
    Very thin and fragile; glassy.

12. *Mitra nodulosa* × 1, 2 views: **KNOBBY MITER**   p. 212
    Elongate; surface with raised granules.

13. *Philine quadrata* × 5, 1 view   p. 227
    Aperture gaping widely.

14. *Melampus coffeus* × 1, 2 views: **COFFEE–BEAN SHELL**

    p. 228
    Shell oval; greenish brown, with paler bands.

15. *Melampus bidentatus* × 1, 2 views: **SALT–MARSH SNAIL**

    p. 228
    Greenish; columella with white patch.

16. *Scaphander punctostriatus* × 1, 2 views: **CANOE SHELL**

    p. 224
    Aperture flaring widely at base.

times forms a bladelike ridge. This species occurs from Cape Hatteras to the West Indies, but is uncommon north of southern Florida.

Family Pyrenidæ

SMALL, FUSIFORM shells, with an outer lip that is commonly thickened in the middle. The inner lip has a small tubercle at its lower end. The shells are sometimes quite colorful, and often shiny. Known as "dove shells," they are confined to warm seas.

Genus *Pyrene* Bolten 1798

4 Species

PYRENE MERCATORIA Linne   (Mottled Dove Shell)

pp. 76, 168

The mottled dove shell is one-half inch in height, and gray in color, clouded and mottled with purplish brown. The sturdy little shell is oval, with five or six whorls and a rather short and blunt spire. The surface bears numerous revolving grooves. The aperture is long and narrow, with the outer lip thickened, particularly in the middle, and strongly toothed within. The inner lip is provided with a series of fine teeth on the lower part. There is a tiny, horny operculum. The mottle dove is abundant on our southern beaches, ranging from Cape Hatteras to the West Indies. It is extremely variable in its coloration, but is always a neat, trim little shell, and a favorite with those who like to make shell novelties. It used to go under the name of *Columbella mercatoria*.

PYRENE RUSTICOIDES Heilprin   (Spotted Dove Shell)

p. 168

The spotted dove shell is usually just a bit bigger than the last species, and its color is yellow or white, with chestnut markings. The shell is oval, with six or seven whorls, a short spire, and a pointed apex. The surface is smooth, often polished. The aperture is long and narrow, the outer lip considerably thickened at the center, and toothed within. There are but few teeth on the lower portion of the inner lip. This species occurs throughout Florida and south to the West Indies. It is more pointed than the mottled dove, and has a much smoother surface. It may be looked for among the grasses in moderately shallow water.

Genus *Anachis* Adams 1853

10 Species and varieties

ANACHIS AVARA Say   p. 168

This shell is usually just under one-half inch in height. Its color is brownish yellow. The shell is thick and strong, with the spire elevated and acute. There are six or seven whorls, sculptured with impressed spiral lines and prominent vertical folds. On the body whorl there are twelve of these folds, which do not descend beyond the middle of the shell, leaving only the revolving lines on that portion. The outer lip is feebly toothed within. Living below the low-tide level, this little gastropod is easily recognized, as there is no other shell on our coast that closely resembles it. It occurs from Cape Cod to Florida, not very commonly in the north but very abundantly in the south.

Genus *Nitidella* Swainson 1840

9 Species and varieties

NITIDELLA CRIBRARIA Lamarck   p. 204

Nearly one-half inch in height, this is a spindle-shaped shell of about seven sloping whorls, with no shoulders, and a spire that is sharply pointed. The aperture is long and narrow, the outer lip thickened and feebly toothed within. The surface of the shell is smooth, and the color is yellowish tan, with a network pattern of brown on the lower half of the body whorl, and a series of brown dots spiraling up the spire. In life the shell has a dark-brown periostracum. The inside of the aperture is bluish white, and shiny. This little mollusk is found on the rocks between tides, in southern Florida and in the West Indies.

Genus *Mitrella* Risso 1826

28 Species and varieties

MITRELLA LUNATA Say   p. 168

This snail is a little mite, scarcely one-fifth of an inch high. It is yellowish tan in color, with reddish brown, crescent-shaped markings. The shell is stout and conical, composed of six whorls separated by shallow sutures. The surface is smooth excepting for a single revolving line below the suture, and a few around the base. The aperture is oval and narrow, the outer lip simple, with a few weak teeth along its inner margin. These gaily colored little snails are found just below the low-water mark, clinging to stones, seaweeds, and dead shells. In the spring they are often found between the tides, on the surface of

MISCELLANEOUS PELECYPODS

1. *Malletia dilatata* × 2, 2 views   p. 10
   Small; plump; squarish end.
2. *Nuculana buccata* × 2, 2 views   p. 7
   Small; pointed end; prominent teeth.
3. *Nuculana concentrica* × 2, 2 views: **CONCENTRIC–LINED
   NUT**   p. 7
   Tiny; white; concentric lines.
4. *Pedalion listeri* × 1, 2 views: **LISTER'S TREE OYSTER**

                                                                       p. 20
   Flattish; pearly inside.
5. *Yoldia myalis* × 2, 2 views   p. 9
   Oval; greenish; prominent teeth.
6. *Malletia obtusa* × 2, 2 views   p. 9
   Brownish; teeth feeble.
7. *Nuculana carpenteri* × 2, 2 views: **NUT SHELL**   p. 8
   Small; curved and narrowed end.
8. *Limopsis aurita* × 2, 2 views: **EARED LIMOPSIS**   p. 17
   See text.
9. *Limopsis sulcata* × 2, 2 views: **SULCATE LIMOPSIS**   p. 17
   Obliquely oval; hairy.
10. *Glycymeris lineata* × 1, 2 views: **LINED BITTERSWEET**

                                                                       p. 18
    Orbicular; plump and solid; curved teeth.
11. *Yoldia arctica* × 2, 1 view: **ARCTIC YOLDIA**   p. 9
    Squarish; gray-green; teeth robust.
12. *Pecten vitreus* × 1, 2 views: **TRANSPARENT SCALLOP**

                                                                       p. 27
    Fragile; translucent.
13. *Pecten striatus* × 1, 2 views: **STRIATE SCALLOP**   p. 27
    Small; pinkish.
14. Surface of *Pecten striatus* × 10
15. *Pinna carnea* × ½, 1 view: **FLESH SEA PEN**   p. 18
    Large; thin and brittle; flesh-colored.
16. *Volsella aborescens* × 1, 2 views: **PAPER MUSSEL**   p. 36
    Elongate; greenish; fragile.
17. *Lima hians* × 1, 2 views: **GAPING FILE**   p. 31
    Oblique; white; ridged.
18. *Limatula hyalina* × 3, 2 views: **LITTLE FILE**   p. 31
    Oblique; white; finely ridged.
19. *Congeria leucopheata* × 1, 2 views: **PLATFORM MUSSEL**

                                                                       p. 39
    Small platform under beaks.
20. *Pecten imbricatus* × 1, 2 views: **LITTLE KNOBBY SCALLOP**

                                                                       p. 26
    Flattish; strongly folded.

Plate 42   193

MISCELLANEOUS PELECYPODS

1. *Cuspidaria rostrata* × 1, 2 views: **ROSTRATE DIPPER** p. 49
   Small; "handle" long and narrow.
2. *Cuspidaria glacialis* × 1, 2 views: **NORTHERN DIPPER**
   Small; "handle" short and stubby.   p. 48
3. *Cuspidaria microrhina* × 1, 1 view: **LONG–HANDLED DIP–**
   PER Larger; "handle" long and thinner.   p. 49
4. *Cuspidaria media* × 2, 2 views: **COMMON DIPPER** p. 49
   See text.
5. *Lyonsia beana* × 1, 2 views: **GAPING LYONSIA** p. 47
   Thin-shelled; glassy.
6. *Cuspidaria glypta* × 5, 1 view: **SCULPTURED DIPPER**
   Tiny; strong radiating folds.   p. 49
7. *Botula fusca* × 1, 2 views: **DUSKY MUSSEL** p. 37
   Short anterior end; beaks hooked.
8. *Lyonsia floridana* × 1, 2 views: **FLORIDA LYONSIA** p. 47
   White; irregular; thin-shelled.
9. Sand grains at margin of *Lyonsia*, × 10
10. *Myonera gigantea* × 1, 1 view: **GIANT MYONERA** p. 50
    See text.
11. *Dacrydium vitreum* × 5, 2 views   p. 37
    Tiny; oblique; brown.
12. *Crenella faba* × 2, 2 views: **LITTLE BEAN MUSSEL** p. 39
    Tiny; oval; fine radiating lines.
13. *Halicardia flexuosa* × 1, 2 views   p. 50
    One-sided; sturdy; chalky-white.
14. *Divaricella quadrisulcata* × 1½, 1 view   p. 57
    White; peculiar "bent" sculpture.
15. Sculpture of *Divaricella*, × 10
16. *Periploma angulifera* × 1½, 2 views: **ANGLED SPOON**
    SHELL White; smooth; projecting tooth under beak. p. 43
17. *Lyonsiella granulifera* × 1½, 2 views   p. 48
    Small; white; obliquely oval.
18. *Lucina leucocyma* × 4, 2 views: **SULCATE LUCINE** p. 56
    Small; sturdy; three-lobed.
19. *Poromya sublevis* × 1, 1 view See text.   p. 48
20. *Volsella opifex* × 1, 2 views: **ARTIST'S MUSSEL** p. 36
    Small; elongate; frayed tip.
21. *Cerastoderma elegantulum* × 1, 2 views: **ELEGANT COCKLE**
    Small, strong radial sculpture.   p. 64
22. *Lithophaga bisulcata* × 1, 2 views: **TWO–FURROWED**
    DATE Elongate; pointed end.   p. 38
23. *Lucina filosa* × 1, 2 views: **NORTHERN LUCINE** p. 57
    Orbicular; compressed; concentric lines.
24. *Gouldia mactracea* × 2, 2 views   p. 53
    Small; triangular; solid.
25. *Panacea arata* × 1, 2 views   p. 42
    Large; thin-shelled; short anterior end.

the sand. This species occurs rather commonly from Cape Cod to Florida, but, like so many of our diminutive mollusks, it is seldom noticed.

Family Nassariidæ

SMALL, CARNIVOROUS snails, present in all seas. The shells are rather stocky, with a pointed spire, a short canal, and commonly a marked columellar callus.

Genus *Nassarius* Dumeril 1805

10 Species and varieties

NASSARIUS TRIVITTATA Say (Little Dog Whelk)
p. 168
The dog whelk is slightly more than one-half inch in height, and white or yellowish white in color, sometimes partly banded with brown. It is a robust little shell, with an acute apex, and six or seven whorls, each flattened a little at the shoulder. A series of spiral grooves cut across a series of beaded lines, giving the surface a "pimpled" appearance. The outer lip is sharp and scalloped by the revolving lines.

Exceedingly abundant between the tides on sandy beaches, the shores in many places are strewn with uncounted hundreds of dead and worn shells of the little dog whelk, many of them perforated with a perfectly round hole, showing that some carnivorous snail, probably one of the *Polinices*, was the cause of its death. This shell may be found from Maine to Florida, but is less abundant south of Cape Hatteras.

NASSARIUS OBSOLETA Say (Basket Shell) p. 168
This species is about four-fifth of an inch high, and varies from dark reddish to purplish black in color. The shell is solid, with six whorls and a moderately elevated spire, rather blunt at the apex. The surface is marked with numerous unequal revolving lines which are crossed by minute growth lines, and more or less oblique folds, especially on the early volutions. The aperture is oval, the outer lip thin and simple, and the inner lip is deeply arched, with a fold at its front. The canal is a mere notch.

This dark and unattractive mollusk is one of the most abundant univalves to be found along the Atlantic coast. It is a scavenger, and a dead fish or a crushed clam thrown into the water will quickly attract hundreds of individuals. These gastropods are to be seen all along the coast, but are especially numerous on muddy shores, where streams render the water somewhat brackish. Adult specimens almost invariably have the apex of the shell more or less corroded.

NASSARIUS VIBEX Say   (Mottled Dog Whelk)   p. 168

The mottled dog whelk is about one-half inch high, and its color is white, variously mottled and marked with chestnut. The shell is short and heavy, with about five whorls. The sutures are well defined, and the apex is pointed. The surface bears strong longitudinal folds, which are crossed by indistinct revolving lines. The aperture is notched at both ends, the outer lip is thick, and there is a heavy patch of enamel on the inner lip.

This is a variable species, and individuals from different beaches are apt to show different color patterns. Long-dead shells tend to lose their chestnut markings, and become a dull white. This little snail occurs from Cape Cod to the West Indies. Very common on Florida shores, it is rare in the north.

NASSARIUS AMBIGUA Montagu   p. 168

This shell is also about one-half inch in height. The predominant color is white, often spotted or banded with brown. The shell is short and stout, with about five whorls, and the apex is sharply pointed. There are about eleven straight ribs on each whorl, extending from suture to suture and crossed by ridges which vary in size. The aperture is circular and quite small, and the outer lip is somewhat thickened. This is an active little gastropod, living from Cape Hatteras to Florida. It is a shallow-water lover, and can generally be collected with little trouble from the tide pools and shallow places along shore, particularly in sandy situations.

Family Buccinidæ

SHELLS GENERALLY large, with few whorls. The aperture is large, and usually notched in front. These are carnivorous snails, occurring in northern seas, all around the world. Some of them are used for food in some countries.

Genus *Buccinum* Linne 1767

24 Species and varieties

BUCCINUM UNDATUM Linne   (Waved Whelk)   p. 169

The waved whelk is three to three and one-half inches in height, and pale reddish or yellowish brown in color. The shell is only moderately solid, regularly convex, and terminates in a sharp apex. The upper whorls are decorated with stout, vertical folds, which become weaker on the later whorls. These folds are crossed by numerous elevated, revolving lines, giving the shell a wavy appearance. The aperture is oval, about one-half the height of the shell, and colored yellow on the inside in fresh specimens. The outer lip is thin and sharp.

This is a circumpolar species, occurring as far south as New Jersey on our coast, It is the edible whelk of Scotland and Ireland. Its station in life is from just below the low-water level to rather deep water. Waved whelks are voracious scavengers and frequently rob the lobster-men of their bait. Most of the empty shells found on the beach are likely to be badly worn or broken, but living specimens are easily obtained by tying up a dead fish in a bag of loose netting and anchoring it among the rocks near the low water line.

BUCCINUM TOTTENI Stimpson   (Totten's Whelk)

p. 169

This is a shell of moderate size, averaging about two inches in height. Its color is yellowish brown, with a thin, pale-yellow periostracum. There are about six rounded whorls, separated by distinct sutures. The shell is sculptured with numerous closely spaced striations that are deeply cut, but the surface is relatively smooth. The aperture is rather broad, and light straw-color within. The outer lip is thin and simple. This species, which lacks the wavy appearance of the last one, is found from Labrador to northern Maine.

BUCCINUM TENUE Gray   p. 169

This shell is brownish yellow in color and about three inches in height. There are about seven convex whorls, with deep sutures. The shell is decorated with numerous vertical folds, but there are no spiral ridges. The aperture is smaller than with most of this genus, but it is still rather large. The outer lip is thin and sharp, and the canal is little more than a notch. There is a horny operculum. This snail lives in rather deep water, from Labrador to the Gulf of Maine.

BUCCINUM ABYSSORUM Verrill   p. 169

This snail averages some two inches in height, and its color, like the others in this group, is dull yellowish brown. The shell is moderately thin, with an acute spire and seven or eight strongly carinated whorls. The volutions are angulated by sharp revolving carinæ, of which there are usually three very prominent ones on the whorls of the spire. The upper one forms a pronounced shoulder. The aperture is rather small, and somewhat semicircular. The outer lip is thin, and sharply angulated by the termination of the revolving lines. The canal is short and nearly straight, and the inner lip bears a small patch of enamel. The periostracum is inconspicuous or lacking. This is a very deep water shell, usually obtained only by dredging. It occurs off the coasts of the Carolinas. There is considerable variation in the relative stoutness and the degree of angulation.

Genus *Pisania* Bivona 1832

2 Species

PISANIA PUSIO Linne   (Pisa Shell)   p. 29
The pisa shell is purplish brown, with bands of irregular dark and light spots often resembling chevrons. Its height is about an inch and a half. The shell is sturdy and strong, with a well-developed spire. The surface is smooth, generally shiny. The aperture is large, nearly half the length of the shell. The outer lip is toothed within, and the inner lip bears a prominent tooth at its upper angle. The canal is short and straight.

The little pisa shell is a smooth, often polished, and strikingly marked species, living on the reefs some little distance from shore. It is taken occasionally in Florida, but is much more abundant in the West Indies.

Genus *Cantharus* Bolten 1798

5 Species

CANTHARUS TINCTUS Conrad   Mottled Spindle
Shell)   p. 168
This shell is about one inch in height. Its color is reddish brown, mottled with white. The shell is solid and conical, with five or six whorls, decorated with low longitudinal folds which are crossed by revolving ridges. The shoulders of the whorls are slightly constricted to form small nodules. The aperture is oval, the outer lip is thickened, and the inner lip is provided with a tooth above. The canal is very short. This is a common snail from South Carolina to Texas. It is a shallow water form, and may be seen clinging to rocks and seaweeds close to shore. It is sometimes confused with the next species, but it is a shorter and more rugged shell.

CANTHARUS CANCELLARIA Conrad   (Cross-barred
Spindle)   p. 168
The cross-barred spindle shell is about an inch and a half in height, and reddish brown in color, more or less mottled with white. The shell is spindle-shaped, moderately elongate, and has a sharp apex. There are five or six whorls, and a sculpture of rather indistinct vertical folds that are crossed by wavy revolving lines. The aperture is elongate-oval, about half as long as the whole shell. The outer lip is thickened, the inner lip has a tooth at its upper angle, and there is a plait at the base of the columella. This colorful little gastropod is found in fairly shallow water, in rocky and stony situations. Its range is about the same as the stouter *tinctus*, from South Carolina to Texas, but this species is most abundant in the Gulf of Mexico.

Genus *Phos* Montfort 1810

2 Species

PHOS CANDEI Orbigny   p. 172

This is a pretty little gastropod, about one inch in height. Its color is pale brownish yellow, sometimes almost white, and sometimes lightly banded. The shell has about nine whorls, an elevated spire, and a sharply pointed apex. There is a sculpture of sharp longitudinal ribs, about fifteen to a whorl, and these are crossed by distinct revolving, threadlike lines, so that the surface has a beaded appearance. The aperture is oval and quite long, and the outer lip is somewhat thickened, and notched at its base. The inner lip bears a few folds at its lower end. The operculum is clawlike. This small snail lives in rather deep water, and occurs from North Carolina to the West Indies.

Family Neptuneidæ

A LARGE FAMILY of medium-to-large gastropods, found in warm, temperate, and polar seas. There is considerable diversity in shape, sculpture, and color within the group.

Genus *Neptunea* Bolten 1798

2 Species

NEPTUNEA DECEMCOSTATA Say   (Ten-ridged Whelk)

p. 76

The ten-ridged whelk is from three to four inches in height, and ashy gray in color, with reddish-brown bands. The shell is large and stout, composed of six or seven whorls, spirally ribbed with raised, reddish-brown keels, or ribs, that give the shell a striking appearance. There are ten of these keels on the body whorl, and three on the upper whorls. The inside of the aperture is white, with the darker ribs showing through faintly.

This is a cold-water snail, living among the rocks well beyond the low-water line. Fishermen frequently bring them up in their nets, and they often manage to get into lobster traps. Most of the specimens found on shore are worn and dull, although after violent storms good examples of this interesting species may be seen on the beach. Quite colorful for a mollusk of northern seas, this snail occurs from Nova Scotia to Massachusetts.

NEPTUNEA DESPECTA TORNATA Gould   p. 172

This is a large and sturdy shell, about four inches in height. Its color is buffy yellow, darker on the revolving ridges. There are about eight convex whorls, producing a rather high, turreted spire. The volutions are encircled with ridges of two sizes. On the body whorl there are five large, elevated, revolving ridges, darker in tone than the rest of the shell, and between them are three small, uncolored ridges. These ridges are not nearly as robust as the keels on the last species. The aperture is large and oval, the inner lip is reflected on the body whorl, and there is a short, strongly curved canal. The inside of the aperture is pale yellow. There is a horny operculum, oval in shape. This species occurs in rather deep water, from Labrador to Massachusetts.

Genus *Volutopsis* Morch 1857

2 Species

VOLUTOPSIS LARGILLIERTI Petit de la Saussaye   (Largilliert's Volutopsis)   p. 201
Formerly called *V. norvegica*, this is a fine large shell, with a length of some four inches. There are about six convex whorls, with well indented sutures, forming a fairly tall spire capped by a rather blunt apex. The surface bears weak revolving lines. The aperture is large, the outer lip sharp and somewhat flaring, and the inner lip is reflected back against the body whorl. The operculum is corneous. The color is yellowish white, with a heavy greenish-brown periostracum. This snail is found in deep water in the North Atlantic, both in this country and in Europe.

Genus *Colus* Bolten 1798

27 Species

COLUS STIMPSONI Morch   (Stimpson's Whelk)   p. 76
This shell averages about three inches in height. Its color is bluish white, with a dark greenish-brown periostracum. The shell is spindle-shaped and elongate, with about eight whorls, the last one equaling two-thirds of the shell. The surface bears distinctly raised, revolving lines beneath the periostracum. The aperture is oval, about half as long as the whole shell, and chalky white within. The canal is moderately long, open, and inclined to turn backward.

This is a deep-water gastropod, found from Labrador to Cape Hatteras. Badly worn shells are occasionally found on northern beaches, but fresh specimens, dredged up by fishermen, are handsome mollusks with their dark-hued, velvety periostracum.

COLUS PYGMÆUS Gould   (Pygmy Whelk)   p. 169
This shell is a little over one inch in height, and grayish in color, also with a greenish periostracum. The shell is stout and fusiform, acutely pointed, and has seven or eight whorls. The surface is marked with numerous revolving lines, but these can

MISCELLANEOUS PELECYPODS

1. *Myrtæa lens* × 2, 1 view   p. 59
   Compressed; white; thick-shelled.

2. *Liocyma fluctuosa* × 1, 2 views   p. 71
   Oval-triangular; concentric lines; shiny.

3. *Rupellaria typicum* × 1, 1 view   p. 73
   Bluntly oval; coarse; dull white.

4. *Tellina angulosa* × 1, 1 view: **ANGLED TELLIN**   p. 75
   Oval; thin; rosy.

5. *Macoma brevifrons* × 1, 1 view   p. 78
   Elongate-oval; low beaks; buffy.

6. *Coralliophaga coralliophaga* × 1, 2 views   p. 73
   Cylindrical; rusty-white.

7. *Parastarte triquetra* × 5, 2 views   p. 72
   Tiny; higher than long; brownish.

8. *Idas argenteus* × 5, 1 view   p. 37
   Small; squarish ends.

9. *Tellina lævigata* × 1, 1 view: **SMOOTH TELLIN**   p. 75
   Roundish; pale orange rays.

10. *Donax fossor* × 1, 1 view   p. 82
    Triangular; prominent posterior slope.

11. *Iphigenia brasiliana* × 1, 1 view   p. 82
    Large; thin-shelled; tan periostracum.

12. *Tellina aurora* × 1, 2 views   p. 78
    Small; oval; polished.

13. *Cyrtodaria siliqua* × ½, 1 view   p. 94
    Elongate-oval; heavy black periostracum.

14. *Semele purpurascens* × 1, 1 view   p. 80
    See text.

15. *Xylophaga dorsalis* × 5, 2 views   p. 100
    Tiny; vertical mid-groove.

16. *Panomya arctica* × 1, 2 views   p. 91
    Large; heavy; chalky-white.

17. *Gastrochæna ovata* × 1, 2 views   p. 94
    Small, gapes widely.

18. *Mesodesma arctata* × 1, 2 views   p. 88
    Short anterior end; deep muscle scars.

19. *Apolymetis intastriata* × 1, 1 view   p. 80
    Large; white; fold at end.

Plate 44   201

MISCELLANEOUS GASTROPODS

1. *Amauropsis islandica* × 1, 1 view   p. 135
    Elongate-globular; thin-shelled.

2. *Solariella infundibulum* × 1, 2 views   p. 115
    Rounded top-shaped; beaded lines.

3. *Bulbus smithii* × 1, 1 view   p. 135
    Globular; flat-topped.

4. *Tegula excavata* × 1, 1 view   p. 112
    Flat bottom; no shoulders; pearly.

5. *Plicifusus kroyeri* × 1, 1 view: **KROYER'S PLICIFUSUS**
                                                       p. 203
    Early whorls fluted.

6. *Volutopsis largillierti* × 1, 1 view: **LARGILLIERT'S VOLU-
    TOPSIS**   p 199
    Weak lines on whorls.

7. *Eudolium crosseanum* × 1, 1 view   p. 176
    Thin-shelled; revolving lines.

8. *Strombus costatus* × ½, 2 views: **RIBBED STROMB** p. 165
    Heavy; thick whitish lip.

9. *Conus verrucosus* × ½, 2 views: **WARTY CONE**   p. 218
    Revolving nodular lines.

10. *Stilifer stimpsoni* × 4, 1 view: **STIMPSON'S STILIFER**
                                                       p. 127
    Tiny; elongate first whorl.

11. First whorl of *Stilifer*, × 10

12. *Calliostoma bairdii* × 1, 1 view: **BAIRD'S TOP SHELL**
                                                       p. 113
    Top-shaped; strongly beaded.

13. *Turritella variegata* × 1, 1 view: **VARIEGATED TURRET**
                                                       p. 151
    High-spired; many whorled; simple aperture.

14. *Velutina lævigata* × 1, 2 views: **SMOOTH VELUTINA** p. 138
    Thin; fragile; large aperture.

only be seen after the periostracum has been removed. The aperture is oval, and the canal rather short.

This snail occurs with the last species, and it may be distinguished from the young of *stimpsoni* by the number of volutions compared with a shell of the same length. The thick velvety periostracum is often corrugated. This species is found from northern New England to Cape Fear.

COLUS SPITZBERGENSIS Reeve   p. 169
This is a rather large species, less fusiform in shape than most of its genus. Its height is nearly four inches, and its color is yellowish gray. There are seven or eight well rounded whorls, decorated with small but prominent revolving ribs, about twenty to a volution. The aperture is nearly circular, the outer lip thin and sharp, and the canal is very short. This is a circumpolar species, living in deep water. It gets down as far as Nova Scotia on our coast, and is well known in Europe.

COLUS LIVIDUS Morch   p. 169
This shell is three or four inches high, and its color is yellowish gray or pale brown. The shell is gracefully elongate, and composed of but five or six whorls. There is a sculpture of revolving ribs, unequal in size. The aperture is long and oval and, with the canal, equals about one-half of the entire shell. The outer lip is thin and somewhat crenulate, and the canal is open and quite long. This species is sculptured much like the last one, but it is much more spindle-shaped, and slender. It occurs from Labrador to Nova Scotia.

COLUS CÆLATULUS Verrill   p. 169
Here we have a small member of this group, about an inch and a half in height. Its color is pinkish gray. The shell is quite sturdy, stoutly fusiform, and has a rather blunt apex. There are about seven whorls, separated by well-impressed sutures. The surface is decorated with stout, vertical ribs, about a dozen to a volution, their upper edge forming a sort of shoulder to the whorl. The outer lip is thin and sharp, and rather evenly rounded. The inner lip is strongly bent and spirally twisted. The canal is fairly long and a little turned to the left. This, too, is a deep-water snail, found from Georges Bank to North Carolina.

COLUS OBESUS Verrill   (Obese Colus)   p. 169
This is a rather fat little fellow, about an inch in height. Its color is pinkish gray. The shell is stoutly fusiform, with four or five whorls, sculptured with many strong vertical ribs and numerous spiral lines. The periostracum bears slender hairs along the spiral lines. The ribs are most pronounced on the convex parts of the whorls. The outer lip is thin, and the inner lip is

strongly excavated in the center. The columella bears a spiral twist. There is a dull greenish-yellow periostracum. This species occurs from Nova Scotia to the Carolinas, living in deep water, well offshore.

COLUS VENTRICOSUS Gray   (Stout Colus)   p. 169
This is a stout, well inflated shell, some two inches in height. Its color is greenish yellow. The shell is swollen, the body whorl comprising most of the shell. There are about five whorls, the sutures well defined. Faint revolving lines decorate the surface, and there is a brownish, velvety periostracum. The aperture is wide, terminating in a short canal, and there is a polished area on the inner lip. This is still another deep-water form, found off the coast of northern Maine.

Genus *Plicifusus* Dall 1902

4 Species

PLICIFUSUS KROYERI Moller   (Kroyer's Plicifusus)

p. 201

This species also inhabits deep water, and is seldom seen in collections. The length is about two inches, and the number of whorls is six. The general shape is quite plump. The early volutions are sculptured with distinct vertical ribs, with the major part of the shell smooth. The aperture is large, the outer lip thin and sharp, and the inner lip partially reflected. There is a short, open canal. The color is yellowish white, with a brownish periostracum. This mollusk has a circumpolar range, and extends part way down on our east coast.

Genus *Busycon* Bolten 1798

8 Species and varieties

BUSYCON CARICUM Gmelin   (Knobbed Pear Conch)

p. 76

This shell is from four to nine inches high. Its color is yellowish gray, with the interior orange red. Juvenile specimens are often streaked with violet. The shell is large and thick, and pear-shaped. There are six whorls, the body whorl large and broad and crowned by a series of blunt nodes, one at each stage of growth. The spire is a low cone, with the series of nodules encircling the shoulders of each whorl. The aperture is long and oval, the canal long and open, and the outer lip thin and sharp. The inner lip is twisted and arched. There is a horny operculum. Normally a right-handed (dextral) shell, left-handed (sinistral) individuals have been collected. The writer was recently privileged to examine such a specimen, collected near Avon, New Jersey.

This is the largest gastropod to be found north of Cape Hat-

VARIOUS MOLLUSKS

1. *Acmæa testudinalis alveus* × 1, 3 views: **TORTOISE–SHELL LIMPET** Small; parallel sides. p. 103
2. *Lucapinella limatula* × 2, 2 views p. 110
   Small; beaded lines; elongate perforation.
3. *Turbonilla elegantula* × 3, 2 views: **ELEGANT TURBONILLA** Tiny; fluted whorls. p. 131
4. *Turbonilla hemphilli* × 3, 2 views: **HEMPHILL'S TURBO–NILLA** Intensely fluted whorls. p. 130
5. *Turbonilla rathbuni* × 3, 1 view: **RATHBUN'S TURBO–NILLA** Larger; fluted whorls. p. 130
6. *Fissurella punctata* × 1, 2 views: **ROCKING–CHAIR LIMPET** p. 107
   Ends higher than sides; perforation in form of cross.
7. *Melanella conoidea* × 1, 2 views Milky-white; shining. p. 126
8. *Pyramidella dolabrata* × 1, 2 views: **OBELISK SHELL** Very glossy; revolving stripes. p. 130
9. *Turbo canaliculatus* × ½, 1 view p. 116
   Large; sturdy; revolving ribs.
10. *Tritonalia cellulosa* × 1, 2 views p. 183
    Small; knobby; forked canal.
11. *Atys sharpi* × 1, 2 views p. 225
    Tiny, white; aperture longer than shell.
12. *Cylichna vortex* × 2, 1 view p. 225
    Tiny; white; aperture narrow at top.
13. *Cerithiella whiteavesii* × 1, 2 views: **WHITEAVES' CERI–THIELLA** Many-whorled; beaded surface. p. 158
14. *Trivia suffusa* × 3, 2 views: **PINK COFFEE BEAN** p. 171
    Small; rosy pink.
15. *Smaragdia viridis* × 4, 2 views: **GREEN NERITINA** p. 120
    Tiny; pale green.
16. *Fusinus amphiurgus* × 1, 1 view p. 209
    Whorls with folds; long canal.
17. *Leucosyrinx subgrundifera* × 1, 1 view p. 219
    Whorls sharply angled; long canal.
18. *Nitidella cribraria* × 3, 1 view p. 191
    Fusiform; outer lip toothed.
19. *Cymatosyrinx moseri* × 1, 1 view p. 219
    Whorls with folds; short canal.
20. *Astræa tuber* × ½, 1 view: **GREEN STAR** p. 117
    Heavy and solid; greenish; pearly within.
21. *Murex antillarum* × ⅔, 1 view: **ANTILLEAN ROCK SHELL** Long canal; spines on varices. p. 182
22. *Gyrineum affine cubaniana* × 1, 1 view: **CUBAN FROG SHELL** Knobby; two varices. p. 179
23. *Hydatina physis* × 1, 2 views p. 226
    Globular; smooth; revolving stripes.
24. *Cypræa cinerea* × 1, 2 views: **GRAY COWRY** p. 170
    Polished; buffy gray with weak bands.

teras. Those found in sheltered bodies of water, such as Long Island Sound, do not reach the maximum size of those living in the open surf. This species occurs on all sorts of bottoms, but perhaps most commonly where the ocean floor is stony. It ranges from Cape Cod to Texas.

Most visitors at the seashore have noted the "egg ribbons" of these snails. The string of curiously flattened, disklike capsules may be picked up alongshore during the summer months. An example is shown on page 173.

On our southern shores we find a variety of this shell that is thick, heavy, and considerably swollen, particularly in the region of the canal. This variety is called *Busycon caricum eliceans* Montfort (p. 173).

BUSYCON CANALICULATUM Linne   (Channeled Pear Conch)   p. 76

This shell gets to be about seven inches in height, and is a pale buffy gray in color, with the interior yellow. The shell is large, pear-shaped, and rather thin, with five or six turreted whorls. The body whorl is very large above, gradually diminishing downward and terminating in a long, nearly straight canal. There is a broad and deep channel at the suture, forming a winding terrace up the spire.

The channeled pear conch is covered in life with a dense, yellowish-brown periostracum that is bristling with stiff hairs. Members of this genus are voracious feeders, overpowering and destroying mollusks nearly as large as themselves. Indian "wampum" was made from the twisted columellas of these snails, cut into elongate beads. This species occurs from Cape Cod to Mexico.

BUSYCON PERVERSUM Linne   (Lightning Conch)
                                                                                                p. 76

The lightning conch is a fawn-colored shell, striped with violet-brown. It grows to a large size, attaining a height of some ten inches or more. The shell is thick and strong, and shaped very much like *B. caricum*, excepting that the spiral turns to the left. In other words, this is a left-handed shell. The spire is rather flatter, and the whole shell more graceful, than *caricum*, so there is little difficulty in distinguishing this species from the rare sinistral examples of the right-handed *caricum*.

This is a southern univalve, and rather brightly colored. It is decorated with more or less zigzag, purplish-brown streaks that give it the popular name of "lightning conch." In old specimens the colors fade, and the shell becomes a dull white. This species is found along our southern coasts, and is rare north of Cape Hatteras.

Just as *Busycon caricum* has a thick and swollen variety, so does *perversum*. It is a dead ringer for the variety *eliceans*, the only difference being that it is left-handed. Its full name is *Busycon perversum kieneri* Philippi.

BUSYCON PYRUM Dillwyn   (Fig Shell)   p. 76
The fig shell is from three to five inches high, and flesh-colored, longitudinally streaked with reddish brown. The shell is large and moderately thin, with a very short spire. The sutures are wide and deeply channeled. The body whorl is large, and it has no spines or knobby shoulders. The aperture is wide, and prolonged into a straight, open canal. The operculum is corneous. The fig shell may be confused with the channeled pear conch, but it can readily be distinguished by its lack of a turreted spire, and its proportionally larger aperture. This snail is common in shallow water, from Cape Hatteras to the Gulf of Mexico, generally preferring sandy situations. In life it possesses a periostracum with numerous short, stiff hairs.

Genus *Melongena* Schumacher 1817
9 Species and varieties

MELONGENA CORONA Gmelin   (Crowned Conch)
p. 105
This species is from two to five inches in height. It is a dark-colored, shiny shell, spirally banded with bluish white, amber, and chocolate in various shades and irregular arrangements. The shell is roughly pear-shaped, with a short spire and a large, inflated body whorl. The shoulders of the last whorl, and about one turn of the preceding whorl, bear a single or double row of short but sharp spines. An additional row or two of blunter spines encircles the base of the shell, or may be lacking. The aperture is oval and wide, the outer lip simple and thin, with a deep notch at its base, and the columella is white and twisted. The operculum is clawlike. This is an active, predatory gastropod, well able to take care of itself in a sea of enemies. It is said that the giant horse conch (*Fasciolaria gigantea*) is the only mollusk capable of overpowering it. It feeds largely upon bivalves, 'coon oysters being perhaps its favorite food, but it does not hesitate to attack and devour *Busycon* and other large snails. Observers have reported what appears to be signs of animal intelligence on the part of these carnivores. They have been noted circling around in back of a resting scallop (*Pecten*) in order to sneak up from the rear, approaching to within pouncing distance before the agile bivalve is aware of danger. The crowned conch prefers muddy and brackish water, and is common throughout Florida. Several varieties have been proposed, based on the shells' difference in height, spine-size, and general sturdiness.

MELONGENA MELONGENA Linne (Brown Crown Conch) p. 105

The brown crown conch is four or five inches in height. Its color is rich brown, with yellow and white bands. The shell is quite solid, composed of about six whorls, with the body whorl making up four-fifths of the total shell, and folding over the short spire, forming a groove at the top of the shell. There are longitudinal ribs on the spire, but not on the body whorl. There may be two or three rows of sharp spines on the last volution, or it may be smooth. The aperture is wide and oval, and the operculum is solid and clawlike.

This shell is common in parts of the West Indies, but is very seldom taken in this country, although it has been reported from the Florida Keys. It is quite variable in the matter of spines, many specimens being found that are entirely smooth.

Family Fasciolariidæ

LARGE SNAILS, with strong, thick, and fusiform shells. The spire is elevated and sharply pointed, and there is no umbilicus. The inner lip is usually decorated with a few oblique plaits. These are predatory mollusks, slow and deliberate in their movements. They are generally distributed in warm seas.

Genus *Fasciolaria* Lamarck 1801

5 Species and varieties

FASCIOLARIA TULIPA Linne (Tulip Shell) p. 105

The tulip shell averages from four to six inches in height. Its color is pale pinkish gray, with interrupted spiral bands of dark brown and many streaks and blotches of reddish brown and amber. The shell is fusiform and moderately high-spired, with eight or nine convex whorls and distinct sutures. The body whorl is swollen, and the surface is fairly smooth, with a few strong wrinkles just below each suture. The aperture is long and oval, the canal short and oblique, and the operculum horny. Although most of the shells of this species that one sees are about five inches in height, eight-inch specimens have been found. The snail occurs from North Carolina to Texas, living from the shore zone out to fairly deep water. It is a sluggish, slow-moving, rather timid snail, but a carnivorous predator nonetheless. Color variations are common, including, rarely, a form that is dark mahogany, with revolving black lines.

FASCIOLARIA DISTANS Lamarck (Banded Tulip Shell) p. 105

The banded tulip shell is a smaller species, averaging about three inches in height. Its color is bluish gray, with longitudinal cloudings of white and revolving narrow lines of rich, dark brown. The shell is gracefully spindle-shaped, or fusiform, with seven or eight whorls. The body whorl is large and swollen, the sutures are distinct, and the surface is smooth, often shiny, with only a few wrinkles at the base. The aperture is long and oval, and the operculum is horny. This shell is very close to the last species, but it is smaller, and its colors are quite different. It occurs from North Carolina to Texas, and is found in much the same situations as its larger relative. The banded tulip is considered by many shell collectors to be one of the prettiest gastropods that can be found commonly on our southern shores, and the writer can think of no more typical shell of Florida beaches.

FASCIOLARIA GIGANTEA Kiener   (Horse Conch) p. 105
This shell has several popular names, among them "horse conch," "pepper conch," and "giant band shell." Its color is brown, with dark revolving lines that become obscure on old shells. The inside of the aperture is orange-red, and the animal itself is brick-red. The shell is ponderous in size and weight. There are about ten whorls, the shoulders of each bearing large but low nodules. The spire is elongate, and there is a sculpture of revolving ridges and strong growth lines. The aperture is wide and oval, and contracted at its base to form a narrow open canal. The columella bears three oblique plaits, and the operculum is corneous. The horse conch is far and away the largest snail to be found in American waters, and it shares with one other species (*Megalotractus auruanus* Linne of Australia) the honor of being the largest univalve in the world. It is at home in moderately shallow water, from North Carolina to Brazil, and while really gigantic specimens are not common, young and medium-sized individuals are relatively abundant. A carnivorous gastropod, this species overpowers and smothers its prey by sheer weight and strength. It is said to be the only mollusk that can successfully cope with the pugnacious crowned conch, *Melongena corona*.

Genus *Latirus* Montfort 1810

6 Species

LATIRUS INFUNDIBULUM Gmelin   (Ridged Latirus)
p. 172
The ridged latirus is about two inches in height. The shell is solid and elongate, with a high spire and a moderately sharp apex. There are about seven whorls, the sutures plainly marked. Each whorl bears about six folds, and the surface is further decorated with prominent spiral ridges, reddish in color. The crests of the folds are commonly worn and somewhat shiny, obliterating the spiral ridges at that point. The canal is quite

long and narrow, the specimen shown on page 172 having an abnormally short canal. There are two or three plaits on the inner lip. This colorful and attractive shell occurs in moderately deep water. Its home is in the West Indies, but it ranges north as far as southern Florida, where it is sometimes collected at the Keys.

Genus *Leucozonia* Gray 1847

2 Species

LEUCOZONIA CINGULIFERA Lamarck   p. 172
This is a rather dark and unattractive snail, some two inches in height. It is brown to brownish black in color, sometimes with a paler band near the base. The shell is rugged and fusiform, with seven or eight whorls, the sutures quite distinct. There are strong tubercles on the body whorl, forming shoulders, and the surface is reticulated by revolving and spiral ridges. The aperture is oval, the outer lip grooved within, and the operculum is horny. This species lives in rather deep water, being found about rocks and coral reefs in from three to six fathoms of water. It occurs from Florida to Texas, as well as in the West Indies.

Genus *Fusinus* Rafinesque 1815

13 Species and varieties

FUSINUS AMPHIURGUS Dall   p. 204
About three-fourths of an inch tall, this is a pretty little shell of some eight whorls, the upper portions of which are sloping. There is a transverse sculpture of very fine lines, plus about a dozen narrow and rounded vertical ribs, with rather wide interspaces. The aperture is rounded, the outer lip wrinkled within, and the canal is slender but prominent. The color is yellowish white. This is a deep-water shell, found in the Gulf of Mexico.

Family Xancidæ

LARGE, THICK, and heavy shells, often ponderous. There are several distinct plaits on the columella. The operculum is clawlike. These snails are all natives of tropical or subtropical seas.

Genus *Xancus* Bolten 1798

1 Species

XANCUS ANGULATUS Solander   (Lamp Shell)   p. 172
The lamp shell is a heavy and ponderous shell, from six to nine inches high. Occasional specimens are a little larger. The color is pale yellowish white, with the interior delicate pink in fresh specimens. There are about six whorls, with prominent knobs

on the shoulders. The lower portion of each volution is marked with revolving lines, especially noticeable on the body whorl. The spire is moderately high, and the apex is bluntly rounded. The aperture is large, descending into a short, open canal, and the inner lip is partly reflected, forming a small umbilicus. The operculum is clawlike, and the columella is strongly plaited. This shell is closely related to the famous "chank shell" of the Indian Ocean, which the Hindus regard as sacred. Formerly known as *Turbinella scolymus* Gmelin, it is found in southern Florida and the West Indies.

Genus *Vasum* Bolten 1798

1 Species

VASUM MURICATUM Born   (Vase Shell)   p. 168
The vase shell averages about three inches in height, and is yellowish brown in color. The shell is rough, heavy, and strong, with a fairly short spire and a bluntly pointed apex. There are six or seven whorls. The surface is sculptured with many revolving ribs and ridges, and there is a series of rather sharp nodes on the shoulders, with sometimes a double row of spine-like nodes on the lower part of the body whorl. The aperture is long, narrowing to the canal, and the inner lip bears transverse folds at the center. There is a heavy brownish periostracum. The vase shell is a solid object, noticeably vase-shaped when inverted, and very knobby or spiny. Juvenile specimens are apt to be much more attractive than mature shells. It lives among the corals and rocks in shallow water, and is a rather common snail in the West Indies. Specimens are occasionally taken in southern Florida.

Family Volutidæ

THE VOLUTES are attractive, colorful shells, and have always been great favorites with collectors, sharing top honors with the cones and the cowries. The shells are more or less vase-shaped, and exhibit a wide assortment of ornamentation and color. The aperture is notched in front, and the columella bears several plaits. The group is noted for having a large, often bulbous initial whorl, or protoconch, at the apex. Volutes are well distributed in tropic seas, living chiefly in deep water.

Genus *Voluta* Linne 1758

1 Species

VOLUTA MUSICA Linne   (Music Volute)   p. 173
The music volute is a handsome species, about three inches in height. Its color is a delicate rosy pink, with brown markings.

The shell is strong and solid, with about six whorls, the large body whorl constituting most of the shell. The sutures are indistinct. The surface is decorated with fine revolving lines in groups of five or six, more or less marked with small dark spots. The outer lip is thick and heavy, and the inner lip has nine or ten robust plaits. The canal is short and strongly twisted. There is a small operculum.

The popular music volute is perhaps as aptly named as any shell from any ocean, for its color pattern of spotted lines certainly does suggest the bars and notes of written music. It is found at the island of Trinidad, and south, living in rather deep water. It is not found in this country, but is shown here as an example of the genus *Voluta*, and because it is so often seen in American collections.

Genus *Scaphella* Swainson 1832

1 Species

SCAPHELLA JUNONIA Hwass   (Juno's Volute)   p. 105

This handsome shell is from three to five inches high. Its color is pinkish white, with slanting rows of squarish, reddish-orange or chocolate spots. The shell is spindle-shaped, the apex pointed. There are about five whorls, with distinct sutures. The body whorl is very large. The surface is smooth, with a very fine periostracum. The aperture is elongate, the outer lip is relatively thin, and the inner lip bears four oblique plaits on its lower part. The canal is short, and there is no operculum.

This is one of the prizes in any shell collection. Good specimens have brought as much as one hundred dollars in the past, when the species was considered extremely rare. It lives among the rocks and corals in deep water, and is still a very uncommon shell on the beach, but sponge fishermen take it occasionally offshore, and every season a few good shells are washed up on shore during storms. Sanibel Island, off the Florida west coast, is one of the best localities for them. The recorded range is from South Carolina to the Gulf of Mexico.

Family Mitridæ

THE MEMBERS of this family, on our shores, are mostly small shells. In the Pacific and Indian Oceans they are much larger. The shell is spindle-shaped, and rather thick and solid, with a sharply pointed spire. The aperture is small and notched in front, and there are several distinct plaits on the columella. The miter shells are confined to warm seas.

Genus *Mitra* Lamarck 1799

20 Species and varieties

MITRA NODULOSA Gmelin   (Knobby Miter)   p 189
The knobby miter, formerly called *granulosa* Lamarck, is about two inches high. Its color is pale brown. The shell is elongate, with a fairly sharp apex and about ten rather flattish but slightly shouldered whorls. The sutures are well impressed. The surface is sculptured with raised granules, produced by longitudinal ribs that are crossed by revolving grooves. There are four plaits on the columella. The aperture is quite short, with a deep notch at its base. This shell may be found from Cape Hatteras to the West Indies.

MITRA SULCATA Gmelin   (Sulcate Miter)   p. 189
This gastropod averages about one inch in height, and is brown in color, sometimes with paler bands. The shell is fusiform, but quite variable in proportions, some specimens being much stouter than others. There are seven or eight whorls, with rather indistinct sutures, and the ornamentation consists of revolving ribs and furrows, faintly marked with vertical lines. The aperture is about half the length of the whole shell, and the columella has several strong, white plications. This species occurs in southern Florida, living among the rocks in moderately shallow water.

Family Marginellidæ

THESE ARE small, porcellaneous, highly polished shells, found on sandy bottoms in warm seas. The spire is short or nearly lacking, and the body whorl is very large. The aperture is narrow and long, the outer lip somewhat thickened, and the columella is plicate.

Genus *Marginella* Lamarck 1801

61 Species and varieties

MARGINELLA APICINA Menke   (Rim Shell)   p. 173
The little rim shell is usually less than one-half inch tall, and its color is ivory or creamy white, sometimes with two or three pale brownish bands. The shell is solid and highly polished, with the spire rather flattened and the body whorl greatly enlarged. The aperture is long and narrow, almost as long as the whole shell, and the outer lip is somewhat thickened. There are four distinct plaits on the inner lip. This is a common little shell in

Florida, and specimens may be found as far north as the Carolinas. It is a very active mollusk, living in bays and shallow water in general. A pure white variety found in southern Florida is known as *M. apicina virginea* Jousseaume.

MARGINELLA GUTTATA Dillwyn   (Spotted Rim Shell)

p. 173

This is a larger shell, about three-fourths of an inch in length. Its color is pale buff, mottled with brown and flecked with white. The brown spots are generally squarish, and quite regularly arranged. The shell is smooth and shiny, with the spire short or flattened. The aperture is elongate, notched at the base, and the outer lip is thickened. Several plaits adorn the columella. Occurring from Cape Hatteras to the West Indies, this snail is larger, but far less abundant, than the last species. It is more colorful, however, with usually three or four dark patches on the rim of the outer lip.

MARGINELLA AVENA Valenciennes   (Banded Rim Shell)

p. 173

This is a pretty little fellow, about one-half inch high. Its color is milky white, with two or three pale, buffy bands encircling the body whorl. The shell is cylindrical in outline, with a short spire, the last volution covering most of the early whorls, and the surface is brightly polished. The outer lip is thin, a little bent in at the center, and the inner lip has four rather strong plaits. This shiny little shell may be found under stones and sponges in shallow water, in southern Florida.

Family Olividæ

COMMONLY CALLED "olive shells," members of this group are more or less cylindrical in shape, with a greatly enlarged body whorl that conceals most of the earlier volutions. The shells are smooth and polished, and often brightly colored. They are widely distributed in warm and tropical seas.

Genus *Oliva* Bruguière 1789

5 Species and varieties

OLIVA SAYANA Ravenel   (Lettered Olive)   p. 29

The lettered olive is a strong and solid shell, about two inches long, sometimes nearly three. Its color is bluish gray, variously marked with chestnut and pink. The shell is quite cylindrical, smooth, and brightly polished. The spire is short, the last whorl covering most of the previous ones. The sutures are deeply incised. The aperture is long, and obliquely notched at its base. The outer lip is thick, and the columella is reflected at its lower end.

This is a gregarious species, generally found in groups in sandy situations where the water is shallow. The pattern of dark markings suggests printed characters, or hieroglyphics, hence the popular name, lettered shell or lettered olive. It used to be known scientifically as *Oliva litterata* Lamarck.

Occurring from North Carolina to Texas and the West Indies, these glossy shells have always been popular for decorative purposes. The coast Indians made necklaces of them long before white men set foot on American shores. There is a pale, yellowish variety that is unspotted. This occurs in southern Florida, and is called *Oliva sayana citrina* Johnson.

OLIVA RETICULARIS Lamarck   (Netted Olive)   p. 173
The netted olive is about one and one-half inches in length. It has a pattern of purplish-brown reticulations on a white or grayish ground. The shell is shaped much like the last species, *sayana*, but it is shorter and relatively stouter. It is just as brilliantly polished. The netted olive has an intricate lacelike pattern of fine lines covering its pale-colored shell. The divergence in color is considerable, ranging from rich brown to pure white, but the netted pattern is rarely lacking. This snail also lives in shallow water, burrowing just beneath the surface of the sand. It occurs in southern Florida.

Genus *Olivella* Swainson 1840
7 Species and varieties

OLIVELLA MUTICA Say   p. 173
This shell is just over one-half inch in height. The usual color is yellowish white, with two or three revolving bands of purplish brown. The coloring, however, is infinitely variable, and specimens will be found that range from nearly white to dark chocolate, with or without bands. The shell is small but solid, with about five whorls and a short, pointed spire. The surface is highly polished. The outer lip is thin and sharp, and there are no plications on the inner lip. This very common little shell is found on mud flats, from North Carolina to the West Indies. Members of the genus *Olivella* may be distinguished from *Oliva* by the larger and more fully developed spire and by the presence of a thin, horny operculum which is lacking in the true *Oliva*.

OLIVELLA JASPIDEA Gmelin   p. 29
This is a shell much like the last species, averaging one-half inch in height. Its color is more constant, however, a pale yellowish gray, with a narrow band of chocolate near the suture. There

are four or five whorls, the body whorl making up most of the shell. The surface is very highly polished. Living in shallow water upon the tidal flats, this species also ranges from North Carolina to the West Indies.

Family Terebridæ

SLENDER, elongate, many-whorled shells, confined to warm and tropic seas. There are no plaits on the columella. Some members of this group are provided with a mild poison, but no American species is dangerous.

Genus *Terebra* Bruguière 1789
14 Species and varieties

TEREBRA DISLOCATA Say   (Little Screw Shell)
pp. 105, 189

This shell, sometimes called the "auger" shell, averages nearly two inches in length, and is ashy gray to pale brown in color. The shell is very elongate, tapering gradually to a fine point. There are about fifteen whorls, with rather indistinct sutures. The surface is decorated with wavy longitudinal folds and fine spiral grooves, and a knobby spiral band encircles the shell just below the suture. The aperture is quite small, there is one distinct twist in the columella, and the yellowish operculum is horny.

The little screw shell is very common on sand bars from Virginia to Texas, and is perhaps most abundant in Florida. The knobby spiral band just under the suture gives the shell the appearance of being composed of alternating large and small whorls. This gastropod occurs very commonly as a Pleistocene fossil in Florida and Bermuda.

TEREBRA PROTEXTA Conrad   (Black Screw Shell)
p. 189

The black screw shell is somewhat smaller than the last species, averaging little more than one inch in height. Its color is not black, but rather a deep chocolate brown. The shell is very long and narrow, with from twelve to fifteen whorls and a very sharply pointed apex. The sutures are indistinct, and each whorl bears a series of sharp-edged folds. The operculum is red brown or claret. This snail occurs from North Carolina to Texas. It, too, is found on sand flats at extreme low tide, although it generally prefers deeper water than *dislocata*. The shell is commonly brown all over, both inside and out, but occasional specimens are paler or whitish on the upper portions of the volutions.

TEREBRA HASTATA Gmelin   p. 189

This shell is about one inch in length. Its color is creamy white, with rather broad bands of pale orange. The shell does not usually taper as regularly as with the others in this group, but commonly remains nearly the same diameter until near the upper end, where it tapers rather abruptly. There are about ten whorls, the sutures quite distinct. The surface of each whorl bears a fine, groovelike plication. The operculum is horny. This is a shiny little shell, occurring in southern Florida and the West Indies.

TEREBRA CINEREA Born   p. 189

This is a larger shell, attaining a length of some two inches. Its color is chocolate brown, with a narrow, whitish band encircling the shell just below the suture. There are about ten whorls which are rather flat, the whole shell tapering very gradually and regularly to a fine point. The surface is shiny, and each volution is sculptured with numerous fine longitudinal grooves. The aperture is small, the outer lip thin and sharp, and the operculum is horny. This species may be collected from Florida to Texas.

Family Conidæ

THIS IS A LARGE family of many-whorled, cone-shaped snails, noted for their variety of colors and patterns. They live among the rocks and corals in tropic seas, only a few members being found on our shores. This group is unusual among mollusks in that some of its members possess poison glands. The venom passes through a tiny duct to the teeth of the radula, and serves to benumb the gastropod's prey. None of the cones living in our waters are known to be dangerous, but certain South Pacific and Indian Ocean species are capable of inflicting serious, and even fatal, wounds.

Genus *Conus* Linne 1758

14 Species and varieties

CONUS SPURIUS ATLANTICUS Clench   (Alphabet
Cone)   p. 104

The alphabet cone is from two to three inches in height, and is the largest cone found on the Atlantic coast. Its color is creamy white, with revolving rows of squarish orange and brown spots and blotches. The shell is smooth and shaped like an inverted cone. There are nine or ten whorls, the first few forming a short, sharp spire. The aperture is long and narrow, and notched at the suture. The outer lip is relatively thin and sharp, and the operculum is horny and very small. This colorful snail, whose irregular markings often do resemble the letters of the alphabet, may be found on both sides of the Florida coast. It is a carnivorous, predatory mollusk, and very active.

For many years this species has been known as *Conus proteus* Hwass, but that name rightfully belongs to an Indian Ocean species which the Florida snail resembles very closely. The typical variety, *Conus spurius*, lives in the West Indies, and its markings are arranged in revolving bands, while the variety *atlanticus*, found in this country, has its markings scattered so that no banding is discernible.

CONUS REGIUS Gmelin   (Cloudy Cone)   p. 104
The cloudy cone is slightly more than two inches in height. Its color is a mottled chocolate brown, sometimes more or less banded. The shell is strong and solid, with from five to eight whorls and a prominent spire with a rounded apex. The surface is sculptured with spiral threads or lines, especially pronounced on the spire. Some of these lines tend to be beaded. The aperture is long and narrow, with the outer lip finely crenulate along the inner margin.

Here the older name, *regius*, must supplant the better-known *nebulosus* Hwass by which this snail has long been listed. It is not a common shell, being found occasionally on the outer reefs from southern Florida to Brazil.

CONUS FLORIDANUS Gabb   (Florida Cone)   p. 104
The Florida cone is about one and one-half inches high, and its color is buffy yellow marked with yellowish brown. The shell has seven or eight whorls, the body whorl constituting most of the shell, as it does with all of the cones. The spire is elevated, and the apex is sharp. The narrow aperture is the length of the shell, excepting the spire, and it is notched at the suture. The inner lip bears weak spiral ridges on its lower margin.

This shell ranges from shallow water to that of moderate depth. It may be found crawling about in rock crevices or under stones, as well as on sand flats. The colors and markings are quite variable, but the shell may be recognized by its sharply pointed apex. It occurs from the Carolinas south.

CONUS CITRINUS Gmelin   (Mouse Cone)   p. 104
The little mouse cone is one inch or so in height, rarely any taller. It used to be called *Conus mus* Hwass. It is a yellowish shell, spotted with reddish brown. There is generally a light-colored, central band. The shape is conical, tapering gradually, with five or six whorls. The spire is low, the apex bluntly rounded, and there may be a row of whitish nodules on the shoulders. The surface bears faint revolving lines. The mouse cone, like its

larger relatives, is an active and predatory snail, overpowering and destroying other mollusks appropriate to its size. It is relatively common in shallow water, and may be found in southern Florida.

CONUS JASPIDEUS Gmelin   (Jasper Cone)   p. 168
The jasper cone averages about one and one-half inches in height, and is white or pale gray in color, banded or mottled with reddish brown. The shell is small, trim, and sturdy. There are about ten whorls, and a rather prominent spire that is finely carinated. There is a sculpture of evenly spaced spiral lines, especially on the lower part of the shell and on the spire. The aperture is elongate, somewhat wider at the base. Like many of the cones, this species will be found in older books under a different name. It used to be called *Conus pealii* Green. It lives in fairly shallow water, and may be collected from the Carolinas to the West Indies.

CONUS JASPIDEUS PYGMÆUS Reeve   (Pygmy Cone)
p. 168
The dwarf cone, or pigmy cone, used to be considered a distinct species, but it is now regarded as merely a variety of *jaspideus*. It is generally less than one inch in height, and its color is bluish white, spotted and clouded with violet-brown. The shell is small, with from seven to ten whorls and an elevated, turreted spire. The shoulders are keeled. The upper part of the body whorl is usually smooth, while the lower part bears revolving grooves. This shell is at home in the Caribbean, and ranges north to the Florida Keys.

CONUS VERRUCOSUS Hwass   (Warty Cone)   p. 201
This is a small cone, seldom more than three-fourths of an inch in height. There are ten or eleven whorls, the shoulders abruptly sloping, so that the elevated spire appears angulated. The aperture is narrow, widening a little at the base. The sculpture consists of rather stout revolving ribs, with regularly spaced beadlike pustules. The color is pinkish gray, more or less blotched with brown. Beach specimens are likely to be pure white. This little cone lives from southern Florida to South America, but it is not very common with us.

Family Turridæ

A LARGE FAMILY of gastropods, many of which are small but highly ornate. The general shape is fusiform, and the outer lip commonly has a slit, or notch. These are lovers of deep water, and occur in warm, temperate, and cold seas.

Genus *Turris* Bolten 1798

4 Species and varieties

TURRIS ALBIDA Dall   p. 172

From two to three inches in height, this is a graceful shell of about eight rounded whorls. The sutures are indistinct. The shell is fusiform, or spindle-shaped, with a high, sharply pointed spire and a rather long, open canal. The outer lip bears a distinct slit (p. 172). The sculpture consists of revolving ridges that are sharp-edged, and oblique striations that decorate the shell between the ridges. The color is pure white. This is a handsome shell, rarely encountered until recently. It lives in deep water off the coast of southern Florida, and the shrimp fisheries have turned up hundreds of them in late years.

Genus *Leucosyrinx* Dall 1889

4 Species

LEUCOSYRINX SUBGRUNDIFERA Dall   p. 204

About one inch in height, this is a graceful shell with an acute spire. There are eight or nine whorls, each sharply angled at the center, so that the spire is pagoda-like. The aperture is small, the outer lip deeply notched at the suture, and there is a long and open canal. Fine lines of growth provide the only sculpture. The color is yellowish white. This snail lives in deep water, from the Carolinas to the Gulf of Mexico.

Genus *Ancistrosyrinx* Dall

2 Species

ANCISTROSYRINX RADIATA Dall   p. 189

This is an oddly beautiful little shell, about one inch in height. As with most of our deep water forms, it is not highly colored, being a drab yellowish gray, or white. There are about six whorls, a moderately high spire, and a long, open canal. Each whorl is sharply angled, producing a steeply sloping shoulder which is ringed with short spines that curve upwards toward the apex. For delicate beauty of form, there are few shells of any size that equal this small univalve. It makes its home in deep water off the Florida coast.

Genus *Cymatosyrinx* Dall 1889

14 Species and varieties

CYMATOSYRINX MOSERI Dall   p. 204

About three-quarters of an inch tall, this is an attractive little snail, found in deep water from North Carolina to the Gulf of Mexico. There are about eight whorls, producing a fairly high spire. The shell is sculptured with curving longitudinal ridges and folds, about eleven to a volution. The aperture is relatively small, the outer lip flaring a little at the base and deeply notched at the suture. The inner lip is rolled back on the body whorl, and the canal is very short. The color of the shell varies from rose-color to pale yellowish white, sometimes with paler bands at the base.

Genus *Clathrodrillia* Dall 1918

23 Species and varieties

CLATHRODRILLIA OSTREARUM Stearns   p. 172
This shell is about one inch high, and reddish brown in color. The shell is strong and solid, composed of from seven to nine whorls, and has a rather blunt apex. There is a prominent elevated ridge following the suture line, effectively separating the volutions. Each whorl bears numerous closely set, longitudinal ribs, which are crossed by revolving, threadlike lines. The aperture is oval, strongly notched at the suture, and opens into a short canal. This species occurs in moderately shallow water, from North Carolina to the Gulf of Mexico.

Genus *Lora* Gistel 1848

32 Species and varieties

LORA HARPULARIA Couthouy   p. 188
This is a buffy or flesh-colored shell, about one-half inch in height. It is stoutly elongate, with from six to eight angled whorls that are flattened somewhat above the angle, forming slightly sloping shoulders. Each volution bears numerous oblique, rounded ribs which are crossed by fine revolving lines. The ribs on the body whorl fade toward the basal margin. The aperture is oval and narrow, the inner lip is smooth, white, and slightly arched, and the canal is a mere notch. This is a deep-water snail, found from Labrador to Rhode Island.

LORA CANCELLATA Mighels & Adams   (Cancelled Lora)
p. 188
This shell is about one-half inch in height, and purplish white in color. The shell is rather slender, with seven or eight turreted whorls. The sculpture consists of numerous vertical ribs, about twenty of the body whorl, which are crossed by raised revolving lines, giving the surface a cancelled appearance. The aperture is small and narrow. Occasionally taken from the stomachs of fishes, this species occurs offshore from Labrador to Massachusetts.

LORA NOBILIS Moller   p. 188
This shell is about five-eighths of an inch in height. Its color is

yellowish white. The shell is stoutly fusiform, with about seven whorls, the shoulders flattened so as to produce a turreted spire. There are about fifteen vertical ribs to a volution, crossed by prominent spiral lines. The aperture is small and narrow, and the canal is short. This shell is found in moderately deep water, from Maine northward.

LORA SCALARIS Moller   p. 188
This snail is about three-fourths of an inch high, and its color is white or grayish yellow. The shell is thin but sturdy, composed of seven or eight well-shouldered whorls. The surface bears rather sharp vertical ribs which become obsolete below the middle of the last whorl. These ribs are crossed by numerous threadlike lines. The outer lip is thin and sharp, and the open canal is very short. This species, too, lives in deep water, from Greenland to Massachusetts.

Genus *Pleurotomella* Verrill 1872
44 Species and varieties

PLEUROTOMELLA PACKHARDII Verrill   p. 188
This is a pale, flesh-colored snail a little more than one-half inch high. The shell is thin and fragile, and somewhat translucent. There are about nine whorls, and an acute, turreted spire. The surface is decorated with prominent rounded, partially oblique ribs, while the spaces between them bear revolving lines. The aperture is rather broad above and elongate below, terminating in a short but wide canal. This species ranges from the Bay of Fundy to Cape Cod. A smaller but sturdier form, with ribs that are less prominent, is found off the Massachusetts coast. This form used to be known as *Pleurotomella saffordi*, but it is now recognized as a variety of *packardii*, listed as *formosa* Jeffreys (p. 188).

PLEUROTOMELLA DALLI Bush   (Dall's Pleurotomella)
                                                              p. 188
This is a pale yellowish white species, not quite one-half inch in height. The shell is turreted and elongate, with about nine convex whorls, the sutures distinctly impressed. Several oblique ribs adorn each volution, about twelve being present on the body whorl. This shell, less fusiform than most of its group, occurs in deep water off the coasts of the Carolinas.

PLEUROTOMELLA JEFFREYSII Verrill   (Jeffrey's Pleurotomella)   p. 188
This is a rather large member of the genus, its height being about one and one-half inches. Its color is a lustrous white. The shell has seven or eight whorls, sharply angled at the sutures, so that the spire is noticeably turreted. The sculpture consists of a row of oblique elongated nodules at the angle of the shoulders, those on the spire most pronounced. These

nodules are continued downward as weak, curved ribs which fade out before reaching the middle of the whorl, leaving much of the shell relatively smooth. This is another deep-water form, found off Chesapeake Bay.

PLEUROTOMELLA EMERTONII Verrill & Smith   p. 188
This shell is about three-fourths of an inch in height, and yellowish white in color, with a thin, glossy, yellowish-green periostracum. The shell is rather stout, with about eight whorls, the body whorl considerably enlarged. The shoulders of the upper whorls bear elongate, oblique nodules, which are almost completely lacking on the body whorl. The aperture is oval and rather broad, and the canal is very short. This snail has been taken in deep water off the coast of Virginia.

PLEUROTOMELLA CURTA Verrill   p. 188
This shell is less than one-half inch in height. Its color is grayish or greenish white, more or less translucent when fresh. There are about five whorls, the body whorl much enlarged, making up nearly three-fourths of the total height. The sutures are deeply impressed, often slightly channelled. The surface is sculptured with fine vertical and revolving lines of nearly equal size. The aperture is moderately large, with a slightly recurved canal at its base. This, too, is a lover of deep water, occurring off the coasts of southern New England.

Genus *Mangilia* Risso 1826

55 Species and varieties

MANGILIA CERINA Kurtz & Stimpson   p. 188
This is a rather elongate little shell, averaging about one-half inch in height. Its color is pale yellowish white. There are about seven whorls with well-developed shoulders and distinct sutures. Each volution bears a few broad, rounded, vertical folds and ribs, crossed by very fine spiral lines. The aperture is small, the canal short, and there is no operculum. This is a mollusk of moderately deep water, living from Massachusetts to Florida. The snail commonly sustains itself upon the surface of the water, with its shell hanging downward.

Genus *Daphnella* Hinds 1844

14 Species

DAPHNELLA LIMACINA Dall   p. 188
This is a shiny little shell about one-half inch high. Its color is ivory-white. The shell has five or six whorls, a pointed apex, and is distinctly fusiform in shape. Most of the surface is smooth and semiglossy, with extremely fine lines of growth,

but there is a prominent row of squarish knobs on the upper part of each volution, making a spiral circuit up the spire. The aperture is moderately long and narrow, and there is a short, open canal. This gastropod prefers rather deep water, and may be found from Martha's Vineyard to the Gulf of Mexico.

Family Cancellariidæ

SMALL BUT SOLID shells, with a striking cross-ribbed sculpture. The aperture is drawn out, with a short canal at the base. The inner lip is strongly plicate, and the outer lip is ribbed within. There is no operculum. These snails are vegetarians, and live in warm seas.

Genus *Cancellaria* Lamarck 1799

5 Species

CANCELLARIA RETICULATA Linne   (Nutmeg Shell)

p. 104

The nutmeg is from one to two inches in height. Its color is bright to pale orange, with weak orange-brown bands. The shell is strong and rugged, with six or seven well-rounded whorls, and the sutures are distinct. The surface is sculptured with longitudinal ribs and revolving lines, producing a network of raised lines over the shell. The aperture is moderately narrow, and the canal is short. The inner lip bears strong oblique plaits.

This snail seldom ventures very far out of its shell, ordinarily showing only the tips of its tentacles as it maneuvers about in shallow water. It feeds upon marine plants, and may be collected from North Carolina to Central America.

Family Acteocinidæ

SMALL "bubble" shells, cylindrical in shape, with or without a short spire. The suture is deeply chaneled, and the inner lip bears one plait. These tiny snails range from cold to tropic seas.

Genus *Acteocina* Gray 1847

6 Species

ACTEOCINA CANALICULATA Say   p. 189

This shell is only about one-fifth of an inch in height. Its color is dull chalky white. The shell is cylindrical, with about five whorls, the summit of each with a shallow, rounded groove. The spire is slightly elevated, but the body whorl constitutes about seven-eighths of the shell. The outer lip arches a little forward, and the inner lip is overspread with a thin plate of enamel and bears one oblique fold near the base.

This diminutive snail is found in shallow water, generally clinging to an old oyster or clam shell. It is also common on decaying, floating timbers. By gathering a few handfuls of the broken fragments found in the coves of a shell beach and running the material through a sieve, one will usually find examples of this gastropod, as well as many other tiny varieties. This species is found all the way from Prince Edward Island to Mexico.

Genus *Retusa* Brown 1827

11 Species and varieties

RETUSA PERTENUIS Mighels   p. 189
This is a very tiny species, less than one-eighth of an inch in height. Its color is dingy white. The surface is smooth, with no traces of sculpture, and there are about four whorls. The early volutions form a spire that is flat and scarcely discernible. The aperture is elongate, narrow above and broad and rounded below. This shell ranges from Greenland to Florida, living in moderately deep water. Specimens are usually obtained from the stomachs of haddock and other bottom-feeding fish.

Family Scaphandridæ

SMALL TO FAIRLY LARGE shells, usually rolled up like a scroll. The shell is thin and brittle. These are carnivorous snails, burrowing in the muds and sands for their prey, chiefly scaphapods.

Genus *Scaphander* Montfort 1810

4 Species

SCAPHANDER PUNCTOSTRIATUS Mighels   (Canoe Shell)
                                                            p. 189
The canoe shell gets to be nearly two inches in length. Its color is pale yellowish brown, and there is a thin, grayish periostracum. The spire is concealed. The aperture is large and white on the inside, and it flares widely at the base. The surface bears numerous fine revolving striæ which are hardly to be seen with the naked eye. This species, which is the largest of the "bubble" type of gastropod on our shores, lives in rather deep water, from the Gulf of St. Lawrence to the West Indies.

Genus *Atys* Montfort 1810

3 Species

ATYS SHARPI Vanatta   p. 204
This is a very thin, elongate-oval shell, about one-fourth of an inch in length. Its shape is somewhat cylindrical, with the two ends tapering. The shell is involute, with the spire concealed. The aperture is long and narrow, with the outer lip rising well above the top of the shell. The surface is smooth, and waxy in appearance, and the color is pure white. This univalve is found in southern Florida.

Genus *Cylichna* Loven 1847

8 Species and varieties

CYLICHNA ALBA Brown   p. 189
This is a rather solid little shell, white in color, with a rusty-brown periostracum. Its length is about one-quarter of an inch, occasional specimens being nearly one-half inch long. The spire is sunken, so that there is a shallow pit at the top of the shell. The surface bears very delicate marks of growth, but the general appearance is smooth and often shiny. This is a rather common little snail along the Atlantic coast from Greenland to North Carolina, and empty shells can often be found in the drift.

CYLICHNA VORTEX Dall   p. 204
Slightly more than one-fourth of an inch tall, this is a bluntly oval little snail, living in deep water. The shell is nearly all body whorl, with a tiny sunken spiral at the top. The aperture is as long as the shell, narrow at the top and wide at the bottom. The shell is thin and brittle, and the color is white or yellowish white. This species ranges from Georges Bank to off Chesapeake Bay.

Family Bullidæ

THESE ARE the true "bubble" shells. They are oval in outline with the shell partly covered by the animal. The aperture is flaring, longer than the body whorl, and is rounded at both ends. Bubble shells are chiefly mollusks of warm seas.

Genus *Bulla* Linne 1758

7 Species

BULLA OCCIDENTALIS Adams   (Bubble Shell)   p. 76
This shell varies considerably in size, its average length being about one inch. The color is pale reddish, mottled with purplish brown, with some individuals showing traces of banding. The

shell is oval and inflated, with the spire depressed. The surface is smooth and polished. The aperture is longer than the shell and is rounded at both ends. There is a reflected white shield at the base of the inner lip. The bubble shells inhabit sandy mud flats, the slimy banks of river mouths, and brackish water in general, concealing themselves in the mud or under seaweeds while the tide is out. They feed upon small mollusks which they swallow whole, crushing them by interior calcareous plates. This snail occurs from Florida to Texas, and is generally plentiful.

Family Hydatinidæ

SHELLS OVAL inflated, and thin in substance. The surface is smooth, and there is no umbilicus. The spire is involute. The animal is large, generally extending beyond the shell. Distributed widely in warm seas.

Genus *Hydatina* Schumacher 1817

1 Species

HYDATINA PHYSIS Linne   p. 204
This is a globose, well-inflated shell, thin and rather fragile, attaining a length of about one and one-half inches. The spire is not concealed as in *Bulla*, but the top of the shell is flattened, and shows a small, tight spiral. The aperture is large and flaring. The surface is smooth and polished, and the color is yellowish gray, with numerous thin, wavy brown lines encircling the shell, every fourth or fifth line heavier than the others. This gastropod enjoys a very wide distribution, being found in Africa, India, Japan, Australia, and in the West Indies and southern Florida.

Family Akeridæ

SMALL AND VERY FRAGILE shells, inhabiting muddy and brackish waters, mainly in warm seas. The animal is too large for its shell, which is partly internal.

Genus *Haminœa* Turton & Kingston 1830

6 Species

HAMINŒA ELEGANS Gray   (Glassy Bubble) pp. 29, 189
This shell is from one-half to three-fourths of an inch high, and greenish yellow in color. The shell is remarkably thin and fragile, and semitransparent. Its shape is much like that of *Bulla*, just described, with the shell a little stouter and more squat. The surface is sculptured with very minute revolving

lines, but the general appearance is smooth and glassy. This species occurs from Florida to Mexico.

Family Philinidæ

SMALL, LOOSELY COILED shells with flaring apertures. The shell is internal, concealed in the mantle of the snail. These are chiefly mollusks of cold seas.

Genus *Philine* Ascanius 1772

13 Species

PHILINE QUADRATA Wood   p. 189
This is a small, delicate, whitish or grayish shell, nearly one-fourth of an inch in height. The thin shell is rather well inflated, with an aperture that gapes widely. The apex is deeply excavated. In these gastropods the shell is not external, but is buried away in the mantle of the animal. This species occurs from New England to Greenland.

Family Siphonaridæ

THESE SNAILS look very much like the limpets, and are sometimes called false limpets. The shell is roughly circular, and conical, with a deep groove on one side that makes a distinct projection on the margin. The animals possess both gills and lungs and spend their time between the tide limits, living a more or less amphibious life, so that they form a connecting link between the purely aquatic snails and the more highly developed air-breathing mollusks.

Genus *Siphonaria* Sowerby 1824

3 Species and varieties

SIPHONARIA ALTERNATA Say   p. 189
This shell is about three-fourths of an inch long, and brown and white in color. The shell is more or less oval in outline, conical, and open at the base. A deep siphonal groove on the right side makes a noticeable projection on that margin, so that the shell is not symmetrical. The surface is decorated with numerous ribs that radiate from the apex, the ribs varying somewhat in size. The apex is commonly eroded. Although these snails look like true limpets, the bulge on the right side immediately distinguishes them. Unlike the lowly limpets, they are equipped with both lungs and gills, and are quite

amphibious. Specimens may be found adhering to stones and dead shells between the tide marks, on Florida beaches.

Family Ellobiidæ

THESE ARE SALT-MARSH snails, spending their time out of the water to a considerable degree. The shell is spiral, and covered with a horny periostracum. The aperture is elongate, with strong folds on the inner lip.

Genus *Melampus* Montfort 1810

5 Species and varieties

MELAMPUS BIDENTATUS Say   (Salt-marsh Snail)

p. 189

This snail is usually less than one-half inch in length. Its color is greenish olive. Young specimens are banded with brown, but old shells are often corroded and coated with a muddy deposit. The shell is oval, thin, and shining when clean. There are about five whorls, the lower one constituting most of the shell, the others flattened to form a short, blunt spire. The aperture is long and narrow, broadest below, and the inner lip is usually covered with white enamel, with two folds crossing the lower part. Deep within the outer lip are several elevated, white, revolving ridges that do not reach the edge of the lip.

This is the commonest salt-marsh snail on the Atlantic coast, and it ranges from Nova Scotia to Texas. It inhabits marshes that are occasionally overflowed by the tide, and is never far from the high-tide mark. When the tide comes in, these snails clamber up the tops of the marsh grass, as if to avoid getting wet for as long as possible.

MELAMPUS COFFEUS Linne   (Coffee-bean Shell) p. 189

The coffee-bean shell is slightly more than one-half inch in length, and is correspondingly stouter than the last species. Its color is pale chocolate, with narrow, creamy bands. The shell is thin but sturdy, with a low spire. There are four or five whorls, and a thin but tough periostracum. The outer lip is thin and sharp, and crenulate within. The columella bears two white folds.

The little coffee-bean is found in southern Florida, where it may be seen on mud flats, or high up on grasses and low shrubs during the full tide. Ordinarily there are three creamy white bands on the body-whorl, but these may be lacking on some specimens. This species, as well as the last one, form a very important food supply for wild ducks.

Glossary of Descriptive Terms

Acuminate — sharply pointed.

Annulated — marked with rings.

Anterior — the forward end of a bivalve shell.

Aperture — the entrance or opening of the shell.

Apex — the tip of the spire in snail shells.

Auriform — shaped like the human ear.

Base — *snails*, the extremity opposite the apex.
    clams, the part of the margin opposite the beaks.

Beaks — the earliest part of a bivalve shell.

Bivalve — a shell with two valves, or shell-parts.

Body-whorl — the last whorl of a snail shell.

Byssus — a series of threadlike filaments that serve to anchor the bivalve to some support.

Callus — a calcareous deposit, such as enamel.

Canal — a tubular prolongation of the lip of the aperture, containing the siphon, in many snails.

Carinate — with a keel-like, elevated ridge.

Columella — the pillar around which the whorls form their spiral circuit.

Coronate — crowned, as in *Melongena*.

Crenulate — notched or scalloped.

Cuspidate — prickly pointed.

Denticulate — toothed.

Dextral — turning from left to right; right-handed.

Diaphanous — transparent, clear.

Discoidal — the whorls being coiled in one plane.

Dorsal — belonging to the back.

Epidermis — *see* Periostracum.

Foot — muscular extension of body used in locomotion.

Gaping — the valves only partially closing.

Gastropod — a snail (or slug).

Genus — a separate group of species, distinguished from all other groups by certain permanent marks called generic characters.

Hinge — where the valves of a bivalve are joined.

Involute — rolled inward from each side, as in *Cyprœa*.

Iridescent — displaying the colors of the rainbow.

Ligament — a cartilage which connects the valves.

Lips — the margins of the aperture of a snail.

Lunule — a depressed area, usually heart-shaped, in front of the umbo, in many clams.

Mantle — a membranous flap or outer covering of the soft parts of a mollusk; it secretes the material that forms the shell.

Nodule — knoblike projection.

Nucleus — the initial whorl.

Operculum — a plate or door which closes the aperture in some snails.

Orbicular — round or circular.

Pallial line — a groove or channel near the inner base of a bivalve shell, where the mantle is made fast to the lower part of the shell.

Pallial sinus — a notch in same.

Penultimate whorl — next to the last whorl.

Periostracum — the non-calcareous covering on many mollusks. Sometimes wrongly called the epidermis.

Plicated — folded or plaited.

Posterior — the backward end of a bivalve shell.

Reticulated — crossed like network.

Rugose — rough or wrinkled.

Septum — diaphragm, platform.

Sinistral — turning from right to left; left-handed.

Siphon — the organ through which water enters or leaves the mantle-cavity.

Species — the subdivision of a genus, distinguished from all others of the genus by certain permanent marks called specific characters.

Spire — the upper whorls, from the apex to the body-whorl.

Striated — marked with fine lines.

Sulcations — furrows, channels.

Suture — the spiral line of the spire, where one volution touches another.

Teeth — the pointed protuberances at the hinge of a bivalve shell; in snails, the toothlike structures in the aperture.

Truncate — having the end cut off squarely.

Turreted — the top of the whorls flattened.

Umbilicus — a small hollow at the base of the body-whorl, visible from below.

Umbones (**umbo**) — the swelling part of bivalve shells, near the beaks.

Undulating — wavelike.

Univalve — a shell composed of a single piece, as a snail.

Varices — prominent raised ribs on surface of a snail shell, caused by a periodic thickening of the lip during rest periods in shell growth.

Whorls — the distinct turns of the spire; also called volutions.

Wing — a more or less triangular projection or expansion of the shell of a bivalve, either in the plane of the hinge-line, or extending above it.

Index

(Numbers in bold-face type indicate illustrations.)